Warm Sands

Warm Sands

Uranium Mill Tailings Policy in the Atomic West

Eric W. Mogren

University of New Mexico Press
Albuquerque

© 2002 by Eric W. Mogren
All rights reserved.
First edition

Library of Congress Cataloging-in-Publication Data:

Mogren, Eric W. (Eric William)
   Warm sands : uranium mill tailings policy in the atomic West /
Eric W. Mogren.—1st ed.
       p.   cm.
   Includes bibliographical references and index.
    ISBN 0-8263-2280-8 (cloth : alk. paper)
  1. Uranium mill tailings—Government policy—United States.
  2. Uranium mill tailings—Environmental aspects—United States.
  3. Radioactive waste disposal—Government policy—United States.
  I. Title.
   TD899.U73 M64 2002
   363.72'89'0560973—dc21
                                                                                     2001003893

For Linda

Contents

Acknowledgments *viii*

Introduction *1*

1. Prologue to Nuclear Danger: The First Atomic Age *17*
2. The Creation of a Government Monopoly *31*
3. The Uranium Boom *51*
4. Warm Water: Tailings and Water Pollution *73*
5. Warm Air: Tailings and Air Pollution *102*
6. Warm Homes: Indoor Tailings Pollution *117*
7. Congress and UMTRCA *145*
8. Closing the Circle *162*

Notes *181*

Bibliography *221*

Index *237*

Illustrations *following page 116*

Acknowledgments

A PROJECT OF THIS SCOPE cannot be completed without the assistance of family members, friends, colleagues, and institutions. Several deserve special recognition for their valuable time, criticism, and suggestions.

When I abandoned my legal career to become a historian, Frederick Allen, Ernest Andrade, Mark Foster, and all my teachers in the History Department at the University of Colorado at Denver encouraged me. Special thanks to my friend and colleague Tom Noel for his early influence and for reminding me about the wonderful breadth of western history.

I am indebted to the faculty at the University of Michigan Department of History who guided me through their remarkably demanding doctoral program. In 1992 I was a displaced westerner among the ivy-shrouded buildings of the central campus. The high plains, lodgepole pine forests, and summer mountain snows of my youth were replaced by the absurdly green and unfamiliar hills of Ann Arbor. When Maris Vinovskis suggested that I explore radioactive waste for my research, I leaped at the idea of pursuing a study that would take me back to familiar territory. Andy Achenbaum offered his time and counsel on this project and all manner of graduate school concerns. Paul Mohai, in the College of Natural Resources, helped me to see the broader implications of my findings and sponsored my Research Partnership Fellowship. Our wonderful community of Michigan graduate student friends studied, raised families, and shared special moments.

I wish to offer special thanks to Sidney Fine, one of the most respected voices in the field of U.S. history, for his faith and support during my years at Michigan. To him I owe my admission to the program, my regular graduate student financial assistance, and my career. His example of professionalism and scholarship instilled in me a profound appreciation of the history discipline. From him I learned the art of teaching, and my dissertation benefited from his sharp analytic skills and graceful, critical editing. I am privileged to have had Sidney as a mentor.

I also gladly acknowledge my debt to the many librarians, archivists, and specialists who helped me to research this book. Those at the following institutions warrant special recognition for their generous service: the Penrose Library, University of Denver; the Marriott Library Special Collections, University of Utah; the New Mexico Environmental Department, Hazardous Waste Bureau, in particular Eloy Montoya; the Colorado Historical Society; the Southwest Research and Information Center; the National Archives and Records Administration; the UMTRA Project Office Archives, Albuquerque; the U.S. Department of Energy Archives; the U.S. Nuclear Regulatory Commission, Public Documents Reading Room; and the Utah Department of Environmental Quality. Without liberal access to their records, this book would not have been possible.

My Northern Illinois University colleagues Michael Gonzalas, Jim Norris, Alan Kulikoff, David Kyvig, and Mary Lincoln read early versions of the book. Other departmental friends there encouraged my scholarship and guided my transition from graduate student to teacher. Thanks also to the university for the Graduate School Research and Artistry Grant that helped me to complete my research.

J. Samuel Walker at the Nuclear Regulatory Commission read several drafts of my manuscript and made insightful critical comments that greatly improved the final version. Other anonymous experts offered equally helpful recommendations. Durwood Ball at the University of New Mexico Press showed early interest in the manuscript, and David Holtby and Evelyn Schlatter steered it through publication hurdles.

I owe my deepest gratitude to my family, whose patience and generosity made this book possible. To my forest ecologist father, who taught me to see the world we look at each day. To my English teacher mother, with her Icelandic passion for literature and history, who taught me the joy of learning. To both of them, for their steadfast faith and support of their younger son. To my older brother, Paul, who has always been an inspiration for me. To my father-in-law, Arnold Gurwitz, whose kindness and unselfish spirit are a guide for my life. To my children, Leif and

Claire, who give me joy. Above all, to my beloved Linda, whose spirit and blessing gave me the courage to change careers, complete graduate school, and finish this book. Throughout the years, we have shared the hard times and rejoiced in our successes together. May it always be so.

EWM
Sycamore, Illinois

Introduction

DURING THE COURSE of two decades, from 1978 to 1998, contractors for the Uranium Mill Tailings Remedial Action (UMTRA) Project removed and secured nearly forty million cubic yards of low-level radioactive uranium reduction mill tailings waste from abandoned mill sites in eleven states and four Indian reservations—enough material to cover 2,300 football fields with ten feet of radioactive sand. They also decontaminated more than five thousand residential, commercial, and public properties polluted with tailings. In addition to these federal efforts, the private uranium industry interred millions more tons of tailings generated by their mill operations. The UMTRA Project was the world's largest material management program designed to shield the public from exposure to potentially hazardous radioactive materials.[1]

On August 26, 1998, the U.S. Department of Energy (DOE) celebrated completion of its extraordinary UMTRA Project with a small party in Grand Junction, Colorado. The modest affair belied the stormy history of uranium tailings pollution. Tailings, the sandy and radioactive effluent of the reduction milling industry, blighted America's atomic achievement. They silently threatened ecosystems in which they were located and jeopardized the health of uranium mill workers and their communities. The tailings crisis also helped to turn Americans against their once-promising Cold War vision of a nation powered by nuclear energy. Although the DOE celebrated the end of the UMTRA Project that warm August day in Grand Junction, many of the guests undoubtedly reflected on the drama of the invidious toxic sand that began at the very dawn of the atomic age.

Nuclear power has a prominent place in America's Cold War dreams. The atom promised, for a while, to remake the world into Utopia. Cold War era political and industrial leaders actively promoted atomic power, tapping into the public's twentieth-century faith that science would solve urgent social and economic problems. One Westinghouse pamphlet proclaimed, "[The atom] will give us all the power we need and more. That's what it's all about. Power seemingly without end. Power to do everything man is destined to do. We have found what might be called perpetual youth."[2] Even the revolutionary Students for a Democratic Society, in its famous 1962 Port Huron Statement, took for granted the truth of the atom's seductive promises and expected nuclear technology to solve intractable human problems. "With nuclear energy," the radicals proclaimed, "whole cities can easily be powered. . . . Our monster cities . . . might now be humanized, broken into smaller communities, powered by nuclear energy, arranged according to community decisions. . . . [A] desire for human fraternity may now result in blueprints of civic paradise."[3] Electrical costs would fall, putting virtually limitless energy within reach of all people. During the mid-1950s, atomic boosters confidently predicted that generating costs for nuclear power plants by 1980 would be half of the already extraordinarily optimistic forecasts for nuclear facilities then under construction. Other dreamers believed U.S. Atomic Energy Commission (AEC) Chairman Lewis Strauss when he suggested to science journalists in 1954 that nuclear power would one day be "too cheap to meter." There appeared to be few limits to the social benefits that would flow from our mastery of the atom.[4]

But the atom had a foreboding countenance, too. It haunted Americans' nightmares with ghastly images of flattened and smoldering Japanese cities, charred bodies, and lingering radiation death. The atom, after all, was born of war and entered America's consciousness as a shocking tool of mass destruction. Popular media and books such as John Hersey's *Hiroshima* brought the bomb's horrors stateside. Children of the Cold War learned to "duck and cover" under their school desks in the event of nuclear attack. Their nervous parents converted basements and backyards into bomb shelters. Black and yellow fallout shelter signs were a familiar sight on public buildings; a few remain today as faded reminders of our dread. The historian Paul Boyer characterized the prevailing emotional climate simply as "primal fear of extinction."[5] The mushroom cloud of an atomic explosion became the icon of modernity. In such a tension-filled postwar America, many people remained wary of the new technology; some feared that humanity had opened Pandora's box by daring to uncover the atomic secrets of existence itself. It is not surprising, then, that Americans found it difficult to distinguish between the benefits and horrors of nuclear energy.

To combat that confusion, atomic scientists, industrial leaders, and government officials labored to educate the nation about the advantages of atomic energy and "sell" it at home and abroad. A new breed of technocratic scientist-statesman inspired Americans to believe that the atom was a powerful tool for peace, profit, and social advancement.[6] The atomic boosters worked to reassure the public that atomic risks had been anticipated and solved. Nuclear energy, they claimed, was destined to rival, perhaps eclipse, fossil fuel as the primary source of the nation's electrical power. Only the most pessimistic critics doubted that it might not be worth the environmental risk. The nation had little to fear and everything to gain from the proper management of the peaceful atom. So forceful was their message and absolute their credibility as masters of the mystical new force that they convinced the nation that nuclear power was essential for the very survival of American civilization.[7]

At the same time, the atomic technocrats and policy makers knew a great deal about the hazards inherent in the practical applications of the new science. They clearly understood that atomic technology posed potential dangers to miners, millers, laboratory and industry technicians, laborers, soldiers, fallout victims—anyone who might come into contact with nuclear materials. Nevertheless, because the primary focus of the nation's atomic energy infrastructure was directed first toward weapons research and production and then toward the speedy privatization of nuclear power, hazard assessment remained a comparatively low priority for the AEC and the Joint (Congressional) Committee on Atomic Energy (JCAE) and its political and industrial allies. "Promotional optimism," noted the economist Steven Cohn, "dominated official hazard assessments."[8] For decades the AEC tolerated "excessive uncertainty" about the risks of nuclear power, which enabled it to portray the atom in the most favorable light. Radiation researchers during the 1950s and 1960s, many of them engaged in "guided research" sponsored by the AEC, did not provide sufficient data to conclusively determine the full, long-term epidemiological consequences of the nation's rush to embrace nuclear power. Their favorable bias toward atomic energy, especially their assessment of the environmental consequences of radiation within relatively short time periods compared with the life span of most radioactive wastes, caused them to regularly underestimate the extent of atomic threats.[9] Both the AEC and the nuclear industry minimized or discredited scientific findings that suggested nuclear risks were greater than advertised. Too often they characterized the discussion about radiation safety as a polemic between reason and fear, rather than an objective debate about the interpretation of scientific data, as a means of disparaging their critics and convincing the public about the safety of the atom. "The burden of proof," Cohn writes, "was placed on proving hazards rather than on proving safety."[10] Finally, these favorable assumptions about atomic energy were institutionalized in regulations that governed the nation's atomic endeavor, including the minimal requirements placed on uranium millers regarding tailings disposal. With such overwhelming support from the public and private nuclear energy establishment, the atom remained the cornerstone of the nation's energy policy for nearly three decades after the end of the war.[11]

Despite such determined efforts to banish the atom's sinister reputation and reassure the public of its benevolence, lingering questions about atomic risk never entirely faded from the national consciousness. Beginning in the environmentally aware 1960s and reaching a crescendo in the 1970s and 1980s, Americans confronted two irreconcilable outlooks on nuclear energy. One view was firmly anchored in the visionary, official, industrial predictions about the benefit and safety of nuclear power. Atomic energy advocates continued to portray it as essential for the well-being of the nation and the world.

The other, more recent outlook raised trenchant and disturbing questions about the economic, human, and ecological consequences of nuclear energy. Nuclear power was never economically competitive with traditional methods of electrical production. Despite the best efforts of the nuclear establishment to link atomic energy with national survival, Americans were not moved to embrace the idyllic vision of atomic energy when it meant higher utility costs. While Washington policy makers and their industrial partners worked to make atomic power a practical reality, the debate over atomic safety transcended the purely technical arena and entered a broader cultural context. The atom's scientific critics found that the confrontational politics of the era offered them a new, friendlier forum in which to contest overly favorable assumptions about atomic safety. Local protests against reactors and other nuclear installations, spearheaded by such grassroots, direct action, antinuclear organizations as the Clamshell Alliance and Ralph Nader's Public Interest Research Group, brought atomic risk to widespread public attention. Industrial whistleblowers were emboldened by the new environmental consciousness. Federal agencies, such as the Department of Labor, challenged the historical AEC and JCAE monopoly over nuclear policy.[12] Troublesome incidents and accidents regularly arose that undermined the Commission's traditional safety claims. With rising confidence, scientists and consumer advocates disputed the prevailing, benign image of the atom, charging instead that nuclear power posed a significant threat to the environment and to public health and was economically unsound. A new breed of experts successfully challenged atomic technocrats' traditional status as stewards of the new technology. By the 1980s Americans came to believe that nuclear power's costs and risks outweighed its benefits. It was a conviction punctuated by two indisputable examples of nuclear carelessness. The first was the nearly disastrous reactor leak at the Three Mile Island nuclear plant in Pennsylvania in 1979. The second was the truly calamitous reactor failure at Chernobyl, Ukraine, in 1986, the nuclear power industry's "first unambiguous catastrophe."[13] These events and other atomic problems, such as the simmering issue of uranium mill tailings, persuaded Americans during the closing decades of the twentieth century to turn their backs on atomic power as an alternative to conventional methods of power production. As Brian Balogh notes, "Ultimately, commercial nuclear power foundered on the twin hurdles of insufficient demand for the product that its experts produced and the loss of the experts' public authority."[14] One key to this atomic reassessment is the interaction between scientific and civilian critics of the AEC, the Commission's response to their objections, and the political climate in which this standoff took place.[15]

While acrimonious discussions of risk came to the forefront of the national debate over nuclear power beginning in the 1960s, the origin of that public reappraisal of atomic safety rests in the atomic pollution issues of the 1940s and 1950s, including the environmental problems associated with uranium milling. Beginning during World War II and continuing at an escalating pace for more than thirty years, public health experts, state governments, and mill communities questioned the safety of the uranium reduction mills that processed domestic ore as part of the federal government's drive to make the United States uranium self-sufficient. Many of their concerns about water purity, air pollution, and mill town contamination were legitimate; some were exaggerated. Accurate or not, however, their increasingly vocal protests invited hostile governmental and public scrutiny of the safety practices of the uranium industry and its governmental champion, the AEC.

Unfortunately for millers, mill communities, and the nation, the uranium industry and the AEC were reluctant to take forceful measures to deal with the early concerns about uranium tailings hazards. Throughout its existence, the Commission, which exercised a legal and virtually monopolistic power over the nation's atomic affairs, was generally reactive, not proactive, in pursuing policies to remedy legitimate health concerns and address fairly fears about uranium mill safety. For decades, the atomic agency interpreted its legal and regulatory mandate in such a way that it largely avoided responsibility for regulating tailings or addressing the full range of environmental threats from the milling process. It made little effort, in its position as the nation's leading authority on atomic matters, to advocate changes in federal atomic laws and policies that might have prevented or remedied mill tailings contamination. Indeed, the agency did all it could to downplay tailings hazards. That legalistic hair-splitting and tentative response to tailings pollution, instead of dynamic leadership from the agency that was charged with safeguarding society from atomic dangers, contributed to the erosion of public faith in the AEC and the nation's atomic endeavor.

The deceptively simple process of uranium reduction milling belies its stormy history. During the halcyon days of America's nuclear power program, uranium ore was mined, processed, fabricated into reactor fuel components, "burned" in fission reactors, and finally disposed of as waste. This step-by-step process is called the "uranium fuel cycle." The milling process is situated between the mine and the refining and fabrication stages in the "front end," or pre-reactor half, of the uranium fuel cycle. Uranium reduction milling is the heavy industry stage of the cycle where uranium-oxide concentrate, called "yellowcake" because of its distinctive canary yellow hue, was extracted from the raw uranium ore. Milling processes varied depending on mill technology and the chemistry of the ore, but generally mountains of incoming ore were pulverized to a sandlike consistency and treated with acid or alkaline solutions that dissolved the uranium and thereby leached it from the ore. Next, the uranium-rich solution was removed from the leaching tank and the uranium solids were precipitated out of the solution, washed, and dried. Finally, workers packed the powdery yellowcake, usually in metal drums, for shipping to facilities where the uranium was further purified and manufactured into reactor fuel rods, weapons, and other items.

The sandlike residue remaining after the mills extracted the uranium from the predominantly low-grade domestic ore was called "tailings," and for every one hundred pounds of ore a mill processed, it deposited more than ninety pounds of tailings, often more than ninety-nine pounds, onto the mill property.[16] The radioactive elements in the tailings were unreclaimed uranium, thorium 230, radium 226, and small amounts of trace elements. Common nonradioactive pollutants in tailings included arsenic, molybdenum, lead, selenium, chloride, manganese, sulfates, and acids or alkaline solvents. Given the extent of uranium production during the Cold War era, it is hardly surprising that a huge amount of these poisonous tailings accumulated at mill sites during the life of the uranium industry. By the 1970s at least 140 million tons of radioactive uranium tailings lay scattered throughout the United States, most of it located in several western states. The tailings piles were enormous. Twenty-seven years of milling near Grants, New Mexico, for example, produced one tailings pile that covered about two hundred fifty acres, rose nearly one hundred feet, and contained perhaps thirty million tons of tailings. As monumental as it was, that pile was hardly unusual for an industry that produced radioactive waste on an unprecedented scale.[17]

Uranium 238, the parent of many naturally occurring radioactive elements, appeared in the earth billions of years ago when gravity coalesced the planet from cosmic debris. Its half-life is 4.51 billion years.[18] During its decay cycle, the lighter decay elements of uranium 238, called "daughters" or "progeny," themselves decay at varying rates, producing their own daughters in a descending cycle until the atom reaches a stable state. Lead 206, for example, is the final stable element of uranium 238. In the process of the uranium 238 radioactive decay sequence, atoms release gamma radiation and high-energy alpha and beta particles.

The biological effects of gamma, alpha, and beta radiation are caused by transferring radiation energy to cellular molecules. Energized cells may be impaired or destroyed. Damaged cells may or may not repair themselves, or they may repair themselves incorrectly. The consequences of radiation exposure depend on the radiation type, the dose, the rate of dose absorption, and the sensitivity of the exposed tissues. External exposures to high radiation doses, from a nuclear explosion, for example, may result in internal body damage or death within a few weeks. Exposure from ingested radiation sources is less likely to produce immediate acute effects but may damage or destroy the organs or tissues in which the radioactive atoms are deposited. Radium, for example, tends to accumulate in bone tissue, thereby causing bone deterioration, leukemia, and other bone-related diseases. The most significant latent effect of any radiation exposure is the increased risk of developing leukemia and cancers of the types that occur naturally in nonexposed populations.

The process of bringing uranium ore to the surface for processing and storing the waste aboveground increased the potential radiation threat from the front end of the uranium fuel cycle. In a typical milling operation in the United States, the incoming ore contained both uranium 238 and uranium 234, which together accounted for only about one-seventh of the total radioactivity in the ore. Because mills usually targeted only uranium for extraction, there was approximately the same amount of radium, thorium, and other radionuclides present in tailings as in the original ore. Significant amounts of radioactive materials remained in the mill waste stream. By 1974 inactive uranium tailings piles in the West contained as much as 15,477 curies of radium, and thousands more were in the piles at operating mills. When radioactive elements leached into the soil and groundwater by surface runoff from the tailings, by percolation through the tailings, or by seepage from tailings containment ponds, they elevated radiation levels in ground and surface waters.[19]

Tailings posed little direct gamma radiation exposure hazard to humans, unless the tailings were used as construction material for buildings, as happened extensively in Grand Junction, Colorado. Nevertheless, gamma radiation is similar to x-rays, and, although not the most hazardous form of radiation associated with uranium milling, it contributed to the overall radiation dosage of anyone exposed to tailings. Gamma radiation is a strong indicator of radioactive materials, and high gamma readings frequently signaled the presence of tailings under or around buildings. The most vexing tailings hazard was the beta and alpha radiation associated with the uranium progeny radium, which has a half-life of sixteen hundred years. Radium decays into gaseous radon 222, with a half-life of 3.8 days. Radon, in turn, decays into four radon daughters, polonium 218 (RaA), lead 214 (RaB), bismuth 214 (RaC), and polonium 214 (RaC¹), all of which have short half-lives ranging from minutes to fractions of a second. Before these daughters decay into the relatively stable lead 210, they emit gamma radiation and alpha and beta particles.

Naturally occurring radium and its decay progeny are present in trace amounts nearly everywhere. When they are deep underground, most radium and radon pose little threat to living things. The small amounts that escape into the biosphere are relatively harmless and constitute part of the planet's natural background radiation. Milling, however, amplified the radiation hazard by concentrating radioactive materials on the earth's surface, increasing their accessibility, and enabling the radiation to enter the biosphere at a greater rate than if the elements had remained locked as subterranean ore.

Although there were many nonradiological industrial hazards in mills, including heavy machinery, organic liquids in solvent extraction facilities, and concentrated acids or alkali in leach plants, it was radiation hazards of radium and its daughters that attracted the most attention and proved to be the greatest source of trouble at the uranium mills. Polonium 218 and polonium 214 are the most hazardous radon daughters because they emit alpha particles that can seriously damage lung tissue and increase the likelihood of contracting pulmonary cancer in people who have inhaled large radon doses. As the radon and its daughters decay inside the lung, they release alpha particles that bombard lung tissue and damage cells, increasing the cancer risk. While scientists still debate the precise levels of radon exposure that cause observable symptoms such as cancer, the link between radon and lung cancer has been beyond dispute for decades. In the European mining districts of Schneeberg and Joachimsthal, since the fifteenth century, middle-aged miners who had worked underground for at least ten years often died from a pulmonary disease that the miners called "Bergkrankheit," or mountain disease. In 1879 doctors diagnosed the disease as lung cancer. Several subsequent studies confirmed that lung cancer was the primary cause of death, and as early as 1924, radon was identified as a significant cause of the disease. By the mid-1930s industrial health experts understood that radioactivity of the ore and the radon gas in the air were the leading cause of lung cancer among the European miners.[20]

Within the uranium mills, the sites of primary radon threat were the ore crushing section, the area where the uranium concentrate was precipitated from the leaching chemicals, and the drying and packaging rooms. In each of these locations, radon daughters were suspended in air. Radon particles tend to attach themselves to solid materials, such as atmospheric dust, water droplets, and cigarette smoke particles. Unprotected workers risked inhaling radon, carrying radon particles home on their clothes, and ingesting it as they ate or drank. Likewise, when radon is concentrated in an enclosed area, like a mine shaft or the room of a building, the likelihood of suffering lung damage and pulmonary cancer is increased as exposure levels rise. Outside, in the immediate vicinity of tailings piles, atmospheric radon levels were higher than background levels as radon diffused through the piles. Wind erosion also blew radioactive particles and dust over mill towns, elevating radon levels beyond natural background levels.

In comparison with other sources of radiation, such as the chain reaction of an atomic weapon, tailings were a fairly low-level radiation hazard. The concern about tailings radiation, however, rested on the long-term consequences of chronic exposure to radioactive elements. That fear was not unfounded. Years of experience with radiation made public health officials sensitive to the possibilities of radiation harm. Observable symptoms of radon exposure can take years to develop, and it was difficult for scientists to predict how much radiation exposure was dangerous or who would contract disease. It was in the context of this early evidence about radium and radon contamination that public health officials became concerned about the possible hazards emanating from uranium mills. They regularly cautioned against placing people in situations where they might absorb low-level radiation over long periods.

For nearly forty years uranium mill tailings provoked periodic public debate. National and regional newspapers, scientific journals, public interest periodicals, and broadcast media regularly carried stories that reflected the growing public and professional awareness and concern about tailings pollution. Yet, despite the prominent role it played in the history of the nation's atomic energy endeavor, to date there is no comprehensive work on the tailings problem and the evolution of uranium tailings policy. Several polemical books, typified by H. Peter Metzger's *The Atomic Establishment* and Ralph Nader's *The Menace of Atomic Energy*, briefly addressed the tailings issue in the broader context of their sweeping antinuclear agenda. Metzger and Nader represent the potboiler-style antinuclear authors who were clearly interested in inflaming public opinion against the nuclear industry and made no effort to explore the mill tailings policy. Two works that provide more scholarly examinations of radioactive waste are *Forevermore: Nuclear Waste in America*, by Donald Bartlett and James Steele, and *Nuclear Imperatives and the Public Trust*, by Luther Carter. These authors, however, provided only a brief glimpse of the tailings problem. The best introduction to uranium mill waste is J. Samuel Walker's *Containing the Atom: Nuclear Regulation in a Changing Environment, 1963–1971*. His short account of the tailings issue provides a good introduction to the federal government's response to the problem. As official historian of the Nuclear Regulatory Commission (NRC), the successor agency to the AEC, Walker focused his brief comments about tailings on government action and the development of tailings policy. Yet his account ends in 1971 and does not cover the significant legislative and public events that presaged the 1978 enactment of the Uranium Mill Tailings Radiation Control Act (UMTRCA), the legal authorization of the UMTRA Project. In addition to these works that include at least some discussion of the mill tailings problem, several books about nuclear waste make passing reference to it, acknowledging mill tailings as a hazard but one that apparently does not require detailed discussion.[21]

The history of uranium reduction milling demands closer analysis. The most appropriate way is to explore the structural factors that drove the formation of tailings policy. Structural variables, such as the legal centralization of authority over atomic energy in the federal government, the de facto autonomy of the AEC and the JCAE, and traditional American democracy, are vital to understanding the evolution of milling policy. Nonelected governmental technocrats, scientists, lawyers, and administrators played a larger and more influential role than did politicians or the public in the uranium mill policy-making process. Governmental organizations and semiautonomous atomic bureaucrats, however, did not function in a predictable, legalistic, and predetermined mechanical way to develop uranium mill policy. Important but decidedly nonstructural factors also influenced mill policy development. Consequently, no state-oriented, structural analysis can, by itself, satisfactorily account for the intangible forces that shaped the policy-making process. State policy formation is not simply the inevitable result of the conduct of abstract government institutions. As Eric Nordlinger has suggested, the state is the "public officials taken all together" that create policy through "all sorts of conflict, competition, and pulling and hauling."[22]

The history of uranium milling shows that several factors, institutional "pulling and hauling," influenced public policy that governed the milling industry. The foundation of mill policy making was the AEC's formal power rooted in the broad legislative authorization of the Atomic Energy Acts. Equally important for the policy-making process, however, were informal forces that reinforced the AEC's legal mandate. Informal influences arose from such sources as agency technocrats' shared assumptions, common goals, monopoly of atomic information, and collective behavior. Another important influence on uranium mill policy, particularly in the later years of the crisis, were scientists inside and outside the governmental atomic infrastructure who leveled sharp and prescient criticism at milling practices. The attitudes of top officials and influential technocrats in leadership positions in the AEC further motivated agency conduct toward the mills. Political goals and assumptions of elected officials swayed agency behavior. Although one hallmark of the nation's atomic policies was their ability to weather changes in elected government and deflect political censure, vocal and strident criticism by federal, state, and local politicians played an important role in shaping uranium milling policy. Similarly, while there was little public disapproval of the AEC or the milling industry at the beginning of the tailings controversy, an increasingly resolute chorus of fearful and angry voices from industry critics was a constant reminder that Commission decisions about uranium milling had real consequences for local communities. The ebb and flow of bureaucratic structure and behavior, even in such an institutionalized agency as the AEC, meant that milling policy was influenced by changing technological, economic, and social trends. Perhaps the most significant influence on tailings policy was the perception of the risks associated with uranium milling. Milling critics predicted grave long-term harm from uranium tailings; federal atomic administrators and the nuclear industry believed that mills posed little threat. In the end, the public sided with the critics. The atomic establishment could not convince mill town residents or the nation that the tailings were a minor threat or successfully convince the public that the benefits of domestic uranium processing outweighed its risks. Therefore, this state-oriented policy history addresses a range of causal variables that "fill in" the gaps that would inevitably arise in a purely structural analysis of uranium milling policy history.

The uranium milling story also reflects both the tension that existed between the AEC and its critics over federal uranium mill policy and the pattern of policy development itself. On the one hand, the federal atomic infrastructure that evolved during the Cold War was extremely powerful. With rare exceptions, the AEC and the JCAE set the nation's atomic agenda, determined the best way of achieving those goals, and carried out the programs. They implemented nuclear policies with a determination that had little tolerance for criticism from either inside or outside the federal government's atomic fraternity. In its response to the tailings issue, a priority for the AEC was to minimize suggestions that uranium milling might be dangerous and reassure the public that hazards from the front end of the uranium fuel cycle were under control. Indeed, atomic technocrats learned very early that "those who control the discourse on risk will most likely control the political battles as well."[23]

On the other hand, as early as the 1950s the AEC came under close scrutiny by state and federal public health experts who believed that the Commission took a lax attitude toward mill pollution. In fact, at the outset of the nation's drive for uranium independence in the early 1950s, the AEC took little notice at all of the long-term environmental, public health, and economic problems associated with the mills. That lack of attention was the result of the legal restraints placed on the AEC by the Atomic Energy Acts, the agency's narrow interpretation of its legal mandate, its belief that tailings were the responsibility of individual states, and its favorable assumptions about the long-term hazards of radiation from tailings. These factors led the Commission to make radiological safety and environmental contamination at uranium mills a low priority. Yet as more and more mill pollution problems came to light, public health officials, local mill town residents, and politicians condemned the AEC for what they believed were harmful mill operations. Equally strong was their criticism of the AEC for what many alleged was a failure to take a leadership role in solving the tailings problems.

This dichotomy resulted in a pattern of tailings policy making: critics identified what they believed to be pollution problem at mills; the AEC responded that the problem was under control or the criticism was unfounded; public and political pressure on the atomic agency grew to a point where the agency could no longer successfully downplay the problem; the AEC sought to assign financial and legal responsibility for the problem to the states, which sought, in turn, to place liability on the federal government; and finally, the AEC and federal and state governments undertook remedial action and reassured the public that the mills were once again safe. The history of tailings, therefore, reflects that action and reaction, cause and effect rhythm that was the hallmark of the uranium mill policy-making process.

Finally, the history of uranium mill tailings policy reveals three fundamental problems that ensured the AEC's efforts to oversee the nation's milling program would fall short. First, the Atomic Energy Act of 1946 and its 1954 amendments created a regulatory scheme that discouraged objectivity in America's atomic establishment about the long-term risks of the nation's atomic program. The acts placed responsibility for developing and promoting atomic energy squarely in the hands of the AEC. Second, at the same time, the law also made the Commission responsible for ensuring the safety of atomic activities. These mutually inconsistent duties would have been a difficult balancing act under the best of circumstances. During the Cold War, when national security interests dominated issues of atomic development, the AEC's promotional efforts clearly outweighed its commitment to environmental health and safety. This problem was compounded by the prevailing assumption of AEC technocrats about their infallibility on nuclear issues, including safety. As David Lilienthal noted in 1963, "The AEC suffered in prestige by [the] unfortunate assumption that, because radiation was its field, it was also an expert on public health."[24] Third, the institutional relationship between the JCAE and the AEC hindered the timely development of effective atomic risk management policies. The authors of the 1946 act envisioned a system of checks and balances that would prevent abuses and safeguard the public in atomic matters. In theory, the AEC would use its technical expertise to formulate new policy proposals and then present them to the JCAE. The Joint Committee, acting as the public's representative and watchdog over the AEC, would then debate the AEC's suggestions and modify, accept, or reject them on behalf of a public that was largely ignorant about atomic science and national security issues. The symmetrical powers of Commission and Committee would control the new technology, but in practice this ideal system fell far short of its goals. By the late 1950s the JCAE had surpassed its regulatory function and sponsored its own atomic policies and directed the AEC to implement its directives. Although the JCAE was sometimes critical of AEC actions, including issues of security, fallout, and the slow pace of nuclear power development, the Committee and the AEC were usually partners, not watchful adversaries, in America's nuclear enterprise. With the AEC predisposed to champion atomic energy rather than regulate it and the JCAE invested in its own promotional efforts, the public and critical scientists were left with few, if any, friendly governmental forums in which to raise troubling questions about atomic safety.

The underlying theme of this book, then, is that the federal government was responsible for the tailings pollution and was the most appropriate entity to undertake tailings management and remediation. For years the AEC asserted that it was not legally obligated to undertake tailings cleanup and made little effort to encourage changes in the laws to increase its responsibilities. Although the 1978 Uranium Mill Tailings Radiation Control Act (UMTRA) finally acknowledged the country's liability for tailings remediation and compelled individual states to shoulder a portion of the cleanup costs, most of the federal government's obligation does not rest on statutory law or legal interpretation. Rather, as Arjun Makhijani, Stephen I. Schwartz, and William J. Weida point out, "[b]y building and operating nuclear production and testing facilities, the government incurred moral obligations both to the workers at those facilities and the communities that coalesced around them."[25] The uranium mills certainly brought short-term economic prosperity to local communities. Yet pollution generated by mills while producing uranium for the nation's nuclear program imposed real long-term burdens on those same communities. By tolerating lax environmental protection standards at the mills and electing to not comprehensively regulate tailings until after millions of tons of them had already been deposited on mill properties, the government shifted part of the costs of national nuclear development onto local communities and natural ecosystems. Aside from passing on economic burdens, the government violated its moral obligation to ensure the safety and health of its citizens. For years the AEC held a de facto monopoly on all aspects of atomic energy. It regularly reassured the public about the safety of the nation's atomic facilities while at the same time working to silence the voices of nuclear experts who raised troubling questions about atomic safety. The historical record leaves little doubt that health experts knew very early the potential hazards associated with low-level radiation from tailings and that government experts and officials regularly underestimated and downplayed those dangers. Even if the precise details of radiation hazards were open to legitimate scientific debate, the very fact that questions existed should have caused the AEC to err on the side of caution, especially in light of the historical trend among radiation health professionals and organizations toward reducing human exposure to radiation.

Uranium mill tailings were only one of several pollution issues that sparked dissent and protest against nuclear power during the last half of the twentieth century. High-level waste disposal generates by far the greatest fear about nuclear energy and remains an issue that overshadows even contemporary discussions of future atomic energy projects. Nuclear safety standards provoked considerable arguments within the scientific community for decades. Reactor safety was always controversial, but especially so after the 1966 incident at the Enrico Fermi Atomic Power Plant reactor in Michigan and the later accidents at Three Mile Island and Chernobyl. Likewise, revelations about lax manufacturing plant safety led to widespread popular outrage. Other issues also generated antagonistic protests. Beginning in the 1950s, uranium mine safety, an issue that highlighted harms similar to those of mill tailings, attracted public and political attention. During the mid-1960s, the U.S. Fish and Wildlife Service suggested that the AEC consider nonradioactive environmental impacts in its licensing process, an issue that quickly erupted into a fierce debate about "thermal pollution" arising from nuclear power facilities.

In retrospect, the tailings controversy was not as socially or politically wrenching for the nation as some of these nuclear problems. Tailings never captured popular attention, for example, to the levels of postreactor waste disposal or reactor safety. Likewise, the cost of tailings cleanup was a fraction of the roughly $365.1 billion spent by the federal government, as of 1996, for nuclear waste management and environmental remediation.[26] The tailings issue is significant for other reasons. Couched as it was in the long-term uncertainty about the potential harm from low-level radiation, it generated sustained regional interest and periodically drew intense national scrutiny, especially during and after the environmentally charged 1960s. Tailings, although the largest case of radioactive waste, were among the last of the long-standing atomic pollution problems to be addressed in a comprehensive manner. Moreover, the monumental physical characteristics of the tailings piles made them a tangible reminder of the environmental costs of the nation's quest for atomic independence. Tailings were not the most prominent nuclear issue, but they were nevertheless a critically important component of the nation's overall nuclear energy policy debate. The increasingly strident scientific and public reaction against tailings hazards reflected the growing environmental consciousness of the postwar period and the evolving public fear about nuclear power. Finally, the remedial action undertaken by states and the federal government serves as an example of the successes, and shortcomings, of the governmental response to an environmental problem it was instrumental in creating.

The implications of this study of uranium milling are far-reaching. Many assume that the acrimonious public debates about nuclear power and the nation's apparent rejection of it as an alternative to other forms of electrical-generating technologies were a sufficient cure for our simplistic postwar assumptions about the benefits and risks of atomic energy. In light of the evidence of the last few decades, that confidence seems misplaced from an environmental policy standpoint. Americans, at heart, still yearn for that Utopia the atom whispered in our dreams; our society today demands power at levels unimagined by even the most optimistic atomic prophet a generation ago. It is probable that nuclear power will again become an important part of our lives in the future. Yet we stubbornly refuse to acknowledge that harm from our historical and current atomic misadventures continue today on an unprecedented scale. Like our parents and grandparents, we remain naively hopeful that science and technology will repair the damage our future atomic labors are certain to inflict on ourselves and our world. The best solution, however, is to look to the past to help us avoid repeating our costly mistakes. Perhaps we may never harmonize our need for energy and our responsibility to our environment; and perhaps we would not be human if did not strive to achieve more tomorrow than we did yesterday. But in light of the accelerating energy demands we put on our world, we must try harder to achieve that balance.

1

Prologue to Nuclear Danger
The First Atomic Age

ON SUN-WARMED GAIOLA hill overlooking the Bay of Naples, the Oxford archaeologist R. T. Gunther discovered in 1907 an exceptional mosaic set in a shallow niche of a crumbled Roman villa. The background was made of bright blue tesserae, depicting the sky, in which hovered a dove rendered in white glass. The foreground was dominated by naturalistic images of plants executed in opaque green glass and supported by a trellis made of yellow tesserae. The Roman mosaic artist had portrayed the vegetation using two hues of yellow-green glass to denote shading on the leaves, and dark blue glass to emphasize the contrast. The Posilipan mosaic, dated to about A.D. 79, was remarkable for its beauty as well as its history, but it also yielded an unexpected surprise: it contained the earliest-known example of glass tinted with uranium.[1]

During the early years of the twentieth century, long before the specter of nuclear war, public ambivalence about atomic power, and divisive policy debates about nuclear waste came to dominate contemporary attitudes toward atomic energy, Americans experienced their own dreams about the atom. Radium, a uranium decay product, launched the first atomic era in the closing years of the nineteenth century. Its ability to release heat spontaneously, the deflection of its radioactive emissions by magnetic fields, and its capacity to luminesce when mixed with other materials attracted widespread scientific and popular attention. Above all, the world was awed by the invisible power of radium to destroy living tissue and end the horror of cancer. Medical and industrial demand for the infinitesimal quantities of radium was so great that it became the world's most valuable commodity. European and American refiners and brokers reaped fortunes from their few grams of the precious element. Radium was the miracle element, symbolic of a new century rich with Progressive optimism and confidence.

Enthusiasm for radium waned, however, as people learned about its sinister qualities. Domestic miners in the ore fields of the remote and inhospitable Colorado Plateau seldom shared in the radium fortunes made by the European and American dealers. Refiners generated large amounts of radioactive uranium by-product that they disposed of indiscriminately. International competition and depression eroded and ultimately destroyed the American radium industry. By the outbreak of World War II, the first atomic age was over. That atomic era, however, served as a foundation for the atomic energy program America developed during and after the war. Reduction milling techniques, perfected first for uranium and then radium extraction, became the prototypes for the postwar uranium industry. Systematic geologic exploration and miners' anecdotes became the road map for prospectors during the uranium boom of the 1950s. The first radium radiotherapeutics served as the basis for modern radiation therapy. The tragic history of medical and occupational uranium and radium poisoning provided the groundwork for modern radiation protection practices.

The splendid Posilipan mosaic proved that the Romans used uranium oxide to manufacture green glass to enhance their mosaic palette. Other cultures valued and used uranium, too. Mining of uranium-bearing uraninite ores in Bohemia and Saxony date back to "beyond statistical record." In the New World, Native Americans, especially those who lived in the Colorado Plateau region, used the yellow and orange-red uranium ore they found along the canyon walls for painting their bodies and pictographs and as pigment for hide garments and moccasins. Uranium first appeared in the contemporary lexicon of metallurgy on September 24, 1789, when Martin Heinrich Klaproth announced to the Prussian Academy of Science that he had isolated "a strange kind of half-metal" from its naturally occurring oxide, pitchblende, or uraninite as it is technically known, and named it in honor of the planet Uranus. Scientists classified the curious element as one of the "rare metals," similar to molybdenum, tungsten, and vanadium.[2]

Klaproth's discovery launched a flurry of scientific investigations of uranium. At least eighteen articles in European science journals between 1789 and 1809 detailed the results of experiments exploring uranium's chemical and metallic properties. Christian Gottlob Gmelin, professor of chemistry at the University of Tübingen, conducted perhaps the most unusual early uranium experiments and published the results of his experiments on the pharmacological effects of uranium salts in 1824. Gmelin reported that uranium compounds he administered to laboratory animals produced severe blood clotting and damage to their internal organs that proved fatal in most instances.[3]

Homeopathic practitioners of the era greeted the scientific evidence about uranium's effects on the human body with keen interest. They claimed that they could cure patients by administering minute doses of materials that in large quantities produced symptoms similar to those of the disease being treated. Homeopathists noted that uranium nitrate damaged the kidneys, resulting in the appearance of sugar in the urine of test subjects. Because sugar in urine is also a symptom of diabetes, homeopathic doctors reasoned that uranium nitrate cured diabetes. By the late 1860s they regularly prescribed uranium solutions for their diabetic patients, a practice that persisted well into the twentieth century despite therapeutic failure and ample evidence about the toxicity of the uranium compounds. One French homeopathic report published in 1930 recommended treating diabetic patients with a *vin urane* cocktail of uranium nitrate, glycerol, and red wine![4]

It was the rediscovery of uranium glass manufacturing techniques during the nineteenth century, however, that propelled uranium from its status as a scientific curiosity and a quack-medicine potion into an important industrial element. In the early 1830s unusual yellow-green glassware attracted enthusiastic industrial and public attention. Bohemian glassmakers obtained those novel hues, as had the Romans two millennia earlier, by adding uranium compounds to their glass formulas. By the early 1840s the French succeeded in copying the Bohemian techniques and were producing large quantities of their own "Bohemian" uranium glass. At the turn of the century, the alluring "canary," "opalescent," "chrysoprase," and "vaseline" varieties of ornamental uranium glass were in great demand among the wealthy on both sides of the Atlantic.[5]

Collectors were enchanted by their uranium knickknacks, but the element had other industrial applications during the nineteenth century that were far more utilitarian than art glass. The ceramic industry used uranium oxide to produce glazes ranging in color from yellow to deep orange. Porcelain and china painters relied on uranium-tinted paints to decorate their products. Uranium oxides colored calico-printing dyes. Uranium was used in ink and occasionally served as a substitute for silver, gold, and platinum in the photographic printing process. Manufacturers used uranium in the mantles of incandescent gas lamps, in cigar lighters, and, later, in the filaments of electric bulbs and glass automobile headlights.[6]

For a century after Klaproth first isolated uranium, it fulfilled important commercial needs, but manufacturers used such small quantities of uranium that producers found it difficult to market more than a few tons of uranium oxides annually. There were no large-scale commercial uses for uranium, while its scarcity kept the price for the metal too high to initiate and sustain significant new industrial demand. Miners recovered relatively small amounts of uranium worldwide, and most of it was found in Europe in conjunction with more commercially valuable metals. The United States, however, had its own uranium ore deposits. The two most important domestic ores were pitchblende, which was the first American uranium-bearing material to attract widespread commercial attention, and carnotite, located primarily on or near the Colorado Plateau. These domestic uranium reserves, first discovered in the late nineteenth century, would eventually help to fuel America's twentieth-century nuclear reactors and atomic weapons. But in the nineteenth-century mines of Saxony and Bohemia, the deep pits of Cornwall, and the mountains and deserts of the American West, the real mining profits were in other metals, and few miners were interested in, or even knew about, uranium.[7]

The absence of any significant commercial demand for American uranium changed around the turn of the century when the Austrian government foreclosed on the bankrupt Bohemian mines in 1899 and halted operations. That closure triggered a brief uranium mining boom in the United States. Optimistic mining speculators believed that a viable domestic uranium industry might be created, and demand for American uranium would increase, if western miners could provide steady supplies of high-quality ore to the European market. Those predictions were bolstered by accounts from European metallurgists that uranium showed promise as an alloy capable of increasing the hardness and tensile strength of steel.[8]

The turmoil at the Austrian mines and favorable reports about uranium steel presaged a bright future for American rare metal producers, but the most significant foundation for an American uranium industry had already been laid by Pierre and Marie Curie. On December 28, 1898, the Curies announced their discovery of the eighty-eighth element, radium, which they had isolated from uranium ore. American miners appeared poised to provide European radium refiners with the raw material from which they could extract the rare and elusive new element.

Domestic miners, however, faced several obstacles in marketing their ore to European uranium and radium refiners. Their high-quality pitchblende deposits were scarce and unpredictable, so prospectors shifted their attention to carnotite, which was easier to find but usually contained far less uranium than pitchblende. Unfortunately, most of the commercially exploitable carnotite ore was located on or near the Colorado Plateau, one of the world's most hostile deserts. Few roads or railroads served the sparsely settled plateau, and at the turn of the century transportation costs for shipping the best-quality ores from its carnotite fields to East Coast dealers were $50 or more per ton, an amount representing at least a third to a half of the market value of top American ores. Under those conditions, only the finest plateau ores could be mined for profit, while the low-grade ores, the kind most commonly found on the plateau, remained unmarketable.[9]

An additional impediment to a profitable domestic uranium industry during the nineteenth century was the absence of any commercial uranium reduction mills in the United States. Continental buyers, accustomed to purchasing rich European and American pitchblende ore, frequently rejected the low-grade Colorado Plateau carnotite ore. American miners, therefore, were obliged to sell their best ore to Europeans, who processed the material themselves and pocketed the profits, while shipping expenses rendered worthless the vast quantities of marginal Plateau ores. A domestic milling industry located in the uranium mining districts could extract and concentrate uranium from the widely available low-grade American carnotite. The enriched mill output would reduce the astronomical transportation costs of Plateau uranium, boost profits, and make domestic ore financially competitive with European uranium. Sensing an opportunity, Charles Poulot, a French chemist, together with his countryman Charles Voilleque, in 1901 built the American Rare Metal Mining and Manufacturing Company reduction mill on the Dolores River, in the McIntyre mining district of Colorado, to process the unprofitable low-grade ores. By constructing their mill near the isolated Plateau mines, they hoped to produce concentrated uranium for continental buyers and eliminate the need to ship raw ores abroad. Using a primitive but effective method of leaching the ore with sulfuric acid and precipitating the residue, the American Rare Metals mill ultimately produced about 15,000 pounds of uranium oxide valued at about $30,000. Imitators built several more reduction mills on the Plateau. The initial success of these mills boosted miners' profits by creating a market for their marginal ores, increasing demand for American uranium, and encouraging new pitchblende and, especially, carnotite prospecting and mining. American uranium production—and profits—increased dramatically. In 1897 Colorado miners shipped abroad only 17 tons of uranium ore they had accumulated over a period of years. Four years later domestic miners produced 375 tons, and by 1902 production jumped tenfold, to 3,810 tons of crude ore, valued at $48,125.[10]

The turn-of-the-century uranium boom was short-lived, however. Although the uranium reduction mills lowered uranium production and transportation costs, world demand for radium and uranium could not sustain prices at a level that would keep the domestic industry competitive with European firms even with the new mills working at full capacity. In 1903 American uranium profits fell to $5,625. By 1905 demand had evaporated and total national production amounted to a paltry $375.00. The reduction mills closed. Colorado Plateau mine owners extracted barely enough material from their locations to satisfy the $100 minimum assessment work required by the mining law to maintain their claims, and then shipped almost none of that ore.[11]

After the crash of the American uranium market, demand for uranium oxides in the glass, ceramics, and dye industries continued to be too low to revive it. Metallurgists and manufacturers concluded that uranium was too difficult and expensive to be used as a steel alloy, especially since satisfactory results could be obtained at lower cost using other elements, including vanadium, a rare metal frequently found in association with uranium in domestic ore. American steel companies began to produce vanadium steel for high-speed tools, gun barrels, springs, and automobile axles. Despite increasing demand for the vanadium, any hope that American miners could cash in on industry's escalating demand for vanadium from domestically mined uranium-vanadium ores was dashed when, in 1906, the American Vanadium Company developed the rich ores of the Minasragra vanadium mine, located near Cerro de Pasco, Peru.[12]

Although the first domestic uranium boom fizzled, rare metal miners of the American West were on the brink of a new mineral rush that would overshadow their first tentative steps to develop the nation's vast radioactive resources. Shortly after the Curies discovered radium, scientists reported that low doses of radiation killed young, rapidly dividing cells but had less damaging effects on mature cells. Within a year of radium's discovery, scientists had investigated its effects on cancerous tumors and had conducted rudimentary radium cancer treatments by 1903. The results of radiation therapy appeared miraculous. Doctors eagerly prescribed the new radiation therapy as an alternative to surgery, especially in parts of the body or for types of malignancies where conventional surgery was difficult or impossible. The public embraced the radium "Curie-therapy" to prevent the horrible disfigurement that was frequently the outcome of early cancer procedures, and survival rates for cancer victims increased dramatically.[13]

Soaring medical demand for radium intensified world hunger for the element. Radium was difficult and expensive to refine, however, because it usually occurred in the ratio of about one part radium to three million parts uranium by weight. Because most American ores usually contained less than 2 percent uranium, thousands of tons of domestic ore needed to be mined and processed to produce even a few grams of radium. Radium's skyrocketing prices caused European producers to reconsider the American ore they had once rejected. Buoyed by this new demand and the prospect of increased profits, domestic rare metal miners returned to the carnotite regions of southwestern Colorado and southeastern Utah to supply the raw material for European radium refiners.[14]

Domestic carnotite ores containing as little as 2 percent uranium oxide, once considered unmarketable, became profitable. But despite increasing demand for American ore, the huge profits made by European radium companies seldom trickled down to the miners. Western miners had little control over the terms of sale for their output. They discarded as much as five tons of low-grade ore for every ton of ore meeting the 2 percent minimum uranium oxide content demanded by most European dealers. Shipping costs for the small quantities of high-quality ore to the Continent continued to erode miners' profits. In 1913, for example, 2 percent ore sold in Europe for about $95 per ton, but total production and transportation costs were about $70 to $75, leaving about $20 per ton profit for the miners. Shipping posed a risk, too, considering all the calamities that could befall a load of ore between the remote mines of the Colorado Plateau and the warehouses of the European radium firms. The shortage of adequate assay facilities on the Colorado Plateau to determine ore quality also hampered miners. They often discarded marketable ore or shipped inferior-quality ore to Europe for which they were not paid. In addition, the shortage of assay equipment meant that miners could not contest the accuracy of assays conducted by the European agents. To make matters worse for the Americans, many European miners enjoyed the commercial advantages of government protection that effectively closed parts of the continental market to local miners altogether. American miners and cancer patients were both victimized by the cold economics of the lucrative radium industry: miners sent their finest ores to European refiners but received little profit in return, while continental refiners sold radium extracted from domestic ore to American hospitals at extravagant prices.[15]

Americans were uncharacteristically slow to meet the challenge of refining radium from their own abundant ore, perhaps because of the earlier financial failures of the reduction mills. But as the Frenchmen Poulot and Voilleque had concluded a decade earlier, American entrepreneurs recognized that low-grade domestic ore could be profitable if it was processed at home. They reasoned that a domestic radium industry, complete with ore reduction mills and radium refineries, might break the European radium monopoly. For miners and investors, local radium refineries meant that radium profits would stay at home. For the public, they meant access to lower-cost medicinal radium supplies.

Despite its slow start, American industry built a series of domestic uranium mills and radium refineries, and at the outbreak of World War I, America was home to three of the world's leading radium companies: Standard Chemical Company of Canonsburg, Pennsylvania; William Lorimer & Sons, of Philadelphia, with its refinery located in Sellersville, Pennsylvania; and the quasi-public National Radium Institute, with refineries located in Denver and near Naturita, Colorado. The American refiners, with access to vast quantities of low-grade domestic ore and efficient refining technologies, quickly surpassed Europe's radium production levels and profits poured into the American companies. The success of these radium refineries signaled the resurgence of the domestic radioactive ore industry.[16]

Demand for American radium plunged in 1914 as European nations turned their attention, and pocketbooks, to the harsh realities of war. But the clouds of conflict hid a silver lining. After the initial price slump, the wartime call for American radium reached a feverish level, because new military applications required increasing amounts of radium at the very time that the war disrupted most European production. Military hospitals, for example, developed innovative radiotherapies for the wounded that reduced scarring. The greatest wartime demand for radium, however, was for luminous paints. Scientists had discovered in about 1903 that if they mixed small amounts of radium with zinc compounds in a bonding material, the resulting paint would glow in the dark. But the luminescent paint remained a curiosity until the war, when the requirement of night warfare created an urgent need for instruments that could be read in the dark. The faint glow from the numbers on the dials made them easier to read but was not bright enough to attract the deadly attention of enemy gunners. To hide soldiers' nocturnal activities, radium luminescent tape guided men through dangerous areas at the front. Soldiers came to rely on luminous watches, gauges, gun sights, spirit levels, and compasses on the battlefield. Between 1917 and 1920 nearly one-half of all domestically produced radium was used for luminous paint alone. American radium firms, with access to secure ore supplies, labor, refining machinery, chemicals, capital, overseas shipping, and the ravenous war-driven demand, dominated the radioactive materials market during the war. They modernized their operations, consolidated existing mines, and located new claims. Production climbed, prices fell, and profits ballooned.[17]

After the war the American firms continued to dominate the radium market. Doctors demanded radium for cancer treatments, and new wartime radiotherapeutic techniques promised peacetime medical benefits. The luminous paint industry consumed large quantities of radium to produce watches, clocks, dials, and such modern "necessities" as glow-in-the-dark light switches and pull-chain illuminators, electrical outlet covers, lock casings, house numbers, doorbell buttons, and crucifixes. The American radium refiners, having expanded and modernized their operations during the war, looked forward to occupying a commanding position in the rare metals market and envisioned themselves as the preeminent producers well into the future.[18]

American producers found, however, that while radium profits steadily increased to stratospheric levels, they could not sell the vast quantities of uranium by-product generated by their radium refining operations. Within a few years after radium production began in earnest, radium manufacturers at home and abroad were generating far more uranium than the market could absorb. The flood of uranium by-product created serious management problems for the rare metal industry. The entire yearly consumption for uranium oxides worldwide in 1914, for example, was about ten thousand pounds. Yet, after only three months of radium processing at the Standard Chemical facility alone, a company representative testified before Congress, "We have on our hands now probably 100,000 pounds [of uranium].... It is stacked up as high as the ceiling. We cannot do a thing with it."[19]

Radium refiners looked desperately for profitable ways to rid themselves of their mounting uranium inventory. Despite the turn-of-the-century findings that uranium was too costly and difficult to use as a steel alloy, the growing stockpiles and plummeting cost of uranium during the two or three years preceding World War I rekindled interest among radium producers for a uranium-steel industry. Even though uranium was available at bargain-basement prices and refiners experimented with new techniques to improve the uranium-steel manufacturing process, the technical obstacles to large-scale uranium-steel production, encountered a dozen years earlier, could not be overcome. All attempts by the radium firms to create a large-scale demand for uranium in the steel industry failed, and their uranium by-product stockpiles continued to soar.[20]

Wartime radium demand exacerbated the uranium oversupply problem, and it became critical for the domestic radium firms to invent ways to dispose of their unmanageable uranium stockpiles. Standard Chemical Company had modest success marketing small quantities of uranium steel for drill bits and high-speed cutting tools. One company even experimented with using it for reversible-pitch airplane propellers. The industrial use of uranium as a coloring agent for glass and ceramics was rekindled briefly after the war, but it could not absorb the mounting supply. Radium refiners grasped at any suggestion that might hold promise of ridding them of their uranium by-product and still allow them to make a profit. Standard Chemical even claimed that low-grade radium residues made excellent plant fertilizers. When the war ended uranium was so plentiful that it was unmarketable. Most radium refiners simply stored their uranium and hoped that it would eventually become profitable. The Radium Luminous Materials Corporation, for example, at first stored its unmarketable uranium oxide in barrels. Later, when the company ran out of storage space, it used the uranium as backfill for its building foundation but also to have it recoverable if the uranium market improved. By 1921 the future of uranium looked so dismal and Radium Luminous had accumulated so much of it that the company simply ran out of storage space and threw it away. Most radium producers followed suit and disposed of their uranium by-product like they would any other refining waste.[21]

While American radium producers sought ways to make money from their uranium by-product, they were content to reap huge radium profits, certain that they would never need uranium revenue to remain solvent. In retrospect, that confidence was misplaced, because the American radium industry's world dominance evaporated in early 1922. Ten years earlier a Belgian mining company, Union Miniere de Haut Katanga, discovered rich pitchblende deposits in the Congo and began secretly mining ore and building refining facilities. Their African ore was the best in the world and as much as twenty-five times richer than most American ore. When Union Miniere finally announced, in 1922, that it was entering the radium market, it had already built the world's largest radium processing plant in Oolen, Belgium, and its exploration reports suggested that their supply of African ore was nearly inexhaustible. The American radium firms were shocked. For more than a year the American radium companies hoped that the Congolese deposits would fail to live up to expectations, but once in full production, the African mines yielded uranium ore at rates even better than predicted.[22]

Overnight Africa became the source of most of the world's radium. The Belgian company undercut American prices by 30 percent or more and radium dropped to its lowest price on record. American radium production virtually ceased within a few months after the Belgians had entered the radium market. Congress showed little enthusiasm for imposing tariffs on imported radium to benefit the American refiners, largely because the medical establishment and the public would have objected to any limitations on cheaper radium. After the dust settled, only one domestic radium plant continued to operate, albeit at greatly reduced capacity. The two largest American radium producers, Standard Chemical Company and Radium Company of Colorado, became marketing agents for the Belgians. Most smaller companies and mines simply closed.[23]

The mining industry is marked by intense cycles of boom and bust, and the radium trade was no exception. During the early 1930s, the Belgian radium industry itself faced stiff competition from Canadian firms, which had their own rich reserves of uranium ore. In 1938 they and the Belgians formed a cartel to split the world market, limit competition, and shore up radium prices in the face of declining world demand. The cartel, however, failed to stabilize the market. Finally, in 1941, the cartel collapsed because the Canadians had amassed a huge and unmarketable inventory of uranium and radium and the Belgian plant closed for lack of raw material when the Germans invaded Belgium.

The sudden failure of the American radium industry, followed by the steady decline of the radium market under Belgian and Canadian dominance, underscores a problem that went beyond simply the cyclical nature of the mining business. What many industry managers failed to appreciate during the interwar years was that the deterioration of the radium industry coincided with the widespread decline in medical and public faith in radium treatments. Frightening revelations about radium poisonings and radium quackery exploded the public's assumptions about radium's safety. By the time the Belgian-Canadian cartel died, the medical community had become wiser and more experienced in its approach to radium therapies, and public confidence in radium, and atomic energy in general, had been badly shaken.[24]

The early radium Curie-therapies presaged the widespread modern use of radioactive materials to treat cancer. But as radium procedures became more common, doctors learned, sometimes tragically, about the dangers of handling radioactive material and administering it to patients.

Many scientists who had experimented with radium died from their exposures, including radium's first victim, Jules Rhens, a radium therapist who died in France in 1905. Some of the earliest public warnings about radiation came from the Deutsche Roentgen Gesellschaft in 1913. This German organization highlighted the dangers of X-rays, made recommendations about the thickness of lead screening to shield workers, and suggested that workers had the right to refuse to undertake radiographic work if the protection was inadequate. In the United States, the Roentgen Society issued similar safety recommendations in November 1915, and over the next thirteen years several other nations introduced further radiation safeguards.[25]

After World War I the medical community's experience with radium caused doctors to temper outlandish prewar predictions about medical miracles to be gained from it. It remained a powerful tool for fighting cancer, worked as a palliative, and helped to reduce deformation associated with tumors, but doctors realized that it had few applications outside the oncological sphere. Perhaps most damaging to radium's public reputation was that it attracted the attention of unscrupulous promoters who exploited the nation's awe, and ignorance, about radium. As uranium had done a generation earlier, radium's mysterious qualities provided quacks with opportunities for unscrupulous profit. The medical community became tragically familiar with the appalling consequences when cancer victims chose radium and thorium patent medicine over legitimate medical treatment.[26]

The event that most galvanized the public's misgivings about radium was the tragedy of the radium luminous dial painters. The work of painting dials with luminous radium paint was a fairly simple one, requiring deftness with the paintbrush. Generally the factories employed teenaged girls. To ensure maximum luminescence, the paint binder was kept to a minimum. Consequently, the paint was thick and the girls found it hard to duplicate the fashionable shaded script; numerals such as 2, 3, and 6 were especially troublesome because the fine curved lines usually ended up too broad. To solve this problem, the girls developed the habit of cleaning their brushes after painting the numbers and then drawing them along the painted lines like an eraser to remove the excess paint and taper the numerals. Many painters found that wiping the brushes on their lips produced the proper point for the job. The girls tipped their brushes up to fifteen times per dial and painted about three hundred dials per day. They usually swallowed the radium paint wiped from their brushes. In the early 1920s a mysterious and fatal bone disease appeared among some former dial painters. In September 1924 evidence confirmed the link between radium and "dial painters' disease." By the late 1920s and early 1930s, medical and popular literature widely reported the lingering deaths of the dial painters and the delayed toxicity of radium.[27]

In 1916 the journal *Radium* reported that "[r]adium has absolutely no toxic effects, it being accepted as harmoniously by the human system as is sunlight by the plant."[28] Within fifteen years of that tragically mistaken conclusion, radium was recognized by the public as a powerful but limited cancer treatment and a fatal poison. In 1929 the *Literary Digest* warned its readers quite simply that "[r]adium deserves to be considered the most dangerous material in the world" and detailed the extreme cautionary practices recommended by the American Medical Association for handling the element.[29] It is remarkable that although scientists had recognized as early as the turn of the century that radiation emitted from radium caused harm to living tissue, they apparently failed to appreciate that radioactive substances continued to emit damaging radiation even after they had been taken into the body. Yet that logical inconsistency, and the tragic consequences flowing from it, was a persistent feature of the nation's experience with radioactive materials before World War II and would continue to influence atomic policy for decades after.[30]

Radium's heyday passed by the late 1930s, and with it the powerful radium companies, monopolies, and cartel. The world's first atomic era, built on the romance of radium and the promise of a future free from the horrors of cancer and disease, was over. As America's rare metal companies disappeared during the interwar era, a few tough, independent "desert rat" miners eked a hard living from the land. The trappings of the once-thriving radium and uranium industry faded into the high mesas and canyons of the Colorado Plateau. Abandoned mines flooded, caved in, or were simply forgotten. Claims lapsed when miners failed to complete their annual assessment work. Mining and milling equipment rusted, and the field offices and mining camps became homes for wild animals. Outside some of the mines, carefully sorted piles of unmarketable uranium ore oxidized to canary yellow. Few people at the time expected that uranium, the once nearly worthless by-product of the radium industry, would become the source of new fortunes in the modern atomic era. "The practice of discarding uranium is very regrettable," prophesied Samuel Lind in 1922, "but . . . [i]t is not at all improbable that a use will some time be discovered for this rare metal which will make it extremely valuable."[31] Lind's prediction was premature, to be sure, and he could not have imagined uranium's future during the second half of the century. Yet within twenty years uranium, a minor industrial metal and the once-worthless by-product of the radium industry, became one of the world's most important strategic commodities, and the Colorado Plateau uranium resources, mines, and mills would eclipse their earlier glory to become central pillars of America's, and the world's, atomic history.

2

The Creation of a Government Monopoly

A HANDFUL OF SCIENTISTS from America's Manhattan Project watched the world's first atomic blast fuse desert sand to glass and light the predawn sky above a remote New Mexico mesa on July 16, 1945. The rest of the world was introduced to the atomic age when President Harry Truman announced that the United States had destroyed the Japanese city of Hiroshima in the blinding flash of light, heat, and concussion of a single atomic weapon at 8:16 on the morning of August 6. The terrible spectacle was repeated three days later at Nagasaki. Japan surrendered on August 10 and World War II ended. The atom made its debut as a tool of mass destruction, shocking even war-weary Americans already accustomed to widespread human and property losses inflicted by the Allies on Germany and Japan during years of conventional warfare.[1]

The fate of the Japanese cities left Americans with a profound anxiety about their atomic future. Three days after the Nagasaki blast, Henry D. Smyth, the official reporter of the Manhattan Project, warned that the technical aspects of atomic development would be overshadowed by political and social questions, the answers to which would "affect all mankind for generations."[2] During the months following the war, however, most Americans were unprepared to provide those answers. The bomb remained a foreboding mystery that few outside the scientific community understood. Government secrecy shrouding the nation's atomic weapons program, a legacy of the wartime security measures, underscored the bomb's menacing qualities. Nevertheless, Americans held general, albeit contradictory, convictions that the nation needed atomic weapons for national defense but that such weapons should never be used again in war. Equally important, the public believed that it should be protected from the threats of radiation, that peaceful nuclear power should be developed free from monopolistic influences, and that the benefits of atomic energy should be distributed fairly. Finally, the nation must develop a comprehensive plan to manage the new destructive technology in a manner consistent with American political traditions and social expectations.[3]

Congress was also struggling over which national values should shape atomic policy. The federal government had traditionally promoted science and technology that contributed to the nation's welfare. Article 1 of the U.S. Constitution mandated federal support for the "progress of science and useful arts" through the patent system. Military research and development always benefited from public support. During the twentieth century especially, government increasingly sponsored laws and assisted research on technologies designed to shield the public from risks of modern industrial advances and natural disasters. Atomic energy, however, challenged these customary roles. Fear that the nation's cities might become smoldering casualties of a future atomic conflict quickly undermined American convictions about the blessings of free market capitalism, the primacy of private property, the power of popular democracy, and the inherent danger of broad federal power. Most members of the postwar 79th Congress, even the most reactionary ones, concluded that the federal government needed to supervise atomic development to maintain America's weapons monopoly, increase the atomic weapons stockpile, and promote atomic energy for public benefit. Disagreements arose about the nature, not the certainty, of that federal control.

Legislative proposals fell into two broad categories—those that favored continued military control of the atom and those that called for civilian oversight. In July 1946, after months of rancorous debate, Congress overwhelming approved the Atomic Energy Act. It embodied a broad outline of civilian control. The heart of the law was a provision that empowered a five-member agency, the Atomic Energy Commission, to regulate military and peaceful uses of atomic energy. In rejecting strong political pressure to keep the new technology under military jurisdiction and placing it civilian hands, Congress made the century's most resounding affirmation of civilian supremacy over the military. The new law, with its complete federal jurisdiction over the revolutionary new technology, departed radically from historical political practices. That it was overwhelmingly approved by a Congress that was generally conservative and preoccupied with dismantling federal bureaucracies created to meet the challenges of the Great Depression and World War II is a testament to the level of national insecurity about the atomic future. Congress, it seemed, had created a plan to safeguard the nation without compromising American political ideals.[4]

Civilian control of the atom, however, did not mean liberal civilian access to atomic secrets. The first section of the act set the restrictive tone for the nation's atomic program, stating that the United States would develop, use, and control atomic energy for the public welfare, "subject at all times to the paramount objective of assuring the common defense and security." Consequently, the AEC was directed to closely regulate fissionable material and atomic devices, as well as the facilities producing and using radioactive materials. The Commission could sponsor private exploration for raw materials, but it was the only entity that could purchase or lawfully own them. Industry needed AEC licenses to transport and process radioactive ore, and the AEC could requisition or condemn supplies of radioactive material or any property containing such material. The agency was empowered to support private research but was also free to conduct research at its own facilities. The act imposed draconian patent restrictions and established new constraints on the free flow of information to private institutions. In essence, it granted the AEC a virtual government monopoly over nearly every aspect of the production, development, and use of atomic energy.[5]

To satisfy fears that it had vested too much control over atomic energy in the hands of the executive branch, and recognizing that secrecy would be impossible if everyone in Congress was involved in atomic policy making, Congress created the Joint Committee on Atomic Energy to oversee the nation's atomic activities and counterbalance the extraordinary powers of the AEC. The JCAE was composed of carefully selected senators and representatives who, in theory, guaranteed that America's nuclear program remained subject to public control.

The JCAE appeared to be the ideal solution to the problem of maintaining legislative supervision over the nation's atomic venture, which was beyond the understanding of most members of Congress, while maintaining conditions of secrecy. The JCAE had several broad functions. It coordinated discussions about atomic matters between the House and the Senate, served as the forum for building compromises between competing versions of proposed atomic-related legislation, and advised the Senate about executive appointments. It provided atomic information to the public on nonclassified aspects of America's atomic program. Finally, it was supposed to be the legislative watchdog over the powerful AEC. This last function was especially important because the mantle of secrecy surrounding nearly all significant projects undertaken by the AEC made traditional political scrutiny of AEC activities nearly impossible. Beyond its official functions, however, the JCAE stood as the symbol of public, civilian control of atomic energy during an era when the complexity of atomic science and secrecy born out of national security fears prevented extensive public participation in atomic policy making.[6]

The centralization of the atomic infrastructure in the hands of the federal government reflected a fundamental public assumption about the new technology. After Hiroshima it was clear that the atom would revolutionize the world. Postwar atomic scientists, policy makers, and popular periodicals took for granted that atomic power would fuel economic and social growth. Atomic energy embodied the era's centralized, corporate ideal of economic development. It was a dominant feature of the postwar "progress package" that claimed that science and technology would solve social problems.[7] Although many claimed that atomic fission technology promised great things for the world, the nation's upbeat assumptions about the new technology were rooted in optimistic, turn-of-the-century Progressive ideals about the fundamental benevolence of the atom.

Nuclear energy was a fashionable topic for the popular press throughout the early twentieth century, and scientists and futurists dreamed of a world transformed by the power of the atom. In a series of 1908 lectures, Frederick Soddy described his vision of the future in which the naturally occurring radioactive decay process would power cities and "transform a desert continent, thaw frozen poles, and make the whole world one smiling Garden of Eden."[8] The science-fiction author H. G. Wells, in his 1914 book, *The World Set Free*, conjured a fictional world of atomic-powered transportation, smokeless industry, and "atomic bombs," all fueled by man-made radioactive elements. The accuracy of his predictions is breathtaking even to modern readers.[9] The very language writers used to describe atomic energy—"philosopher's stone," "elixir of life," "Garden of Eden"—was filled with poetic romance and a sense of magical well-being.[10]

Physics appeared to support such lavish forecasts. Scientists speculated about the energy necessary to hold atoms together and what might happen if, somehow, atoms could be broken apart and the energy released. In 1902 the Curies concluded that the atoms of a radioactive material act individually as constant sources of energy. The following year Soddy and Sir Ernest Rutherford speculated "that the energy latent in the atom must be enormous compared to that rendered free in ordinary chemical change.... [T]here is no reason to assume that this enormous store of energy is possessed by the radio-active elements alone."[11] Albert Einstein published his theory of relativity in 1905, which detailed his concept of the equivalence of matter and energy. Among other theoretical insights, Einstein's suggested that the conversion of mass to energy would release tremendous amounts of power.[12]

Influential scientists accepted Einstein's hypothesis and the theory of atomic power but were skeptical that a practical method for releasing atomic energy from matter would ever be found. The obstacles appeared insurmountable.[13] Rutherford, then president of the influential British Association for the Advancement of Science, reported to his colleagues in 1924 that although matter was the storehouse of energy, in most instances atomic energy would never be available to man, and the hope that the energy contained in the elements could be utilized in the future was unrealizable.[14]

Such pessimism about the prospects for atomic power did not deter physicists from working to uncover the secrets of the atom. In 1930 Ernest Lawrence built the first cyclotron, a device capable of accelerating charged atomic particles at high velocity, that became an indispensable atomic research tool. Two years later James Chadwick discovered the neutron. Even conservative theorists began to believe that nuclear power might be possible. The most promising hypothesis for inducing an atom to release energy focused on "fission," that is, the process of using neutrons from one atom to bombard the nucleus of another atom. Such a collision would smash the atom's nucleus, releasing large amounts of energy, while neutrons from the demolished nucleus would, in turn, shatter the nuclei of neighboring atoms to produce more energy and fragments, and so on in an escalating "chain reaction." In late 1938 Otto Hahn and Fritz Strassman, working at the Kaiser Wilhelm Institute for Chemistry in Berlin, succeeded in splitting a uranium atom. In the United States, similar results were obtained in January 1939 by John Dunning and Enrico Fermi.[15] Word of these successful fission experiments spread rapidly among the physics community, and by the end of 1939, researchers had published nearly one hundred scientific articles detailing their investigations into atomic fission.[16]

The discovery of atomic fission also rekindled futuristic fantasies for an atomic utopia that had fired public imagination during the early years of the twentieth century. Once the practical scientific and technical barriers to atomic energy had been overcome, a new generation of popular writers trumpeted the atom as an energy source that would revolutionize humanity and create an "unparalleled period of richness and opportunity for all."[17] They predicted that the new world prosperity made possible by atomic energy would render privilege and class distinctions relics of history. War would become obsolete because the economic stresses that cause it would no longer exist. Uranium would supplant gold as the standard of value, people would fly in propeller-less airplanes fifty miles above the earth, food would be grown underground throughout the year, and ships fueled by a mere pint of atoms would traverse the globe. Torrents of electricity for home owner consumption would gush from household atomic "black boxes."[18]

Despite the promise for a world reborn by atomic energy, forecasts for a peaceful atomic-powered future appeared remote during the late 1930s. The optimistic dreams of a nuclear utopia were eclipsed by economic turbulence, social disillusionment, and political upheaval that rocked the nations of the world. International tensions even imposed themselves on the usually nonpartisan and scholarly domain of nuclear science. Several prominent European atomic scientists, persecuted by their governments, fled their countries for safety in Britain and the United States. Within a year after scientists had split the atom, the world itself split apart as nations plunged into World War II. Atomic scientists like the Italian Fermi, the Hungarian Leo Szilard, and even the German Einstein himself, well aware of the possibility for atomic weapons and mindful that Germans had been on the cutting edge of fission experimentation, persuaded the United States to initiate its own military atomic program. During the conflict, the Allied and Axis powers directed their atomic science programs toward military applications, not to developing technologies that would uplift humanity.

Americans were proud of the scientific and technological achievement represented by the atomic bomb and thankful that the new weapon had ended the war so abruptly. The nation, however, had little time to rejoice over its Pacific victory. The fate of the Japanese cities brought with it a profound anxiety about what the new technology would mean for the future. The public also confronted new ideas and images about atomic energy that reinforced its apprehension about the new technology. The press introduced Americans to a new atomic vocabulary, such as "fission," "nucleus," "fallout," and "chain reaction," that sounded strange, scientific, and vaguely threatening. Most Americans had never witnessed an atomic blast firsthand, but widely circulated photographs and films of postwar atomic tests conducted at Bikini Island in the South Pacific on July 1, 1946, familiarized Americans with the visual images of atomic warfare.

Books like Hersey's *Hiroshima* portrayed in graphic detail the effect of an atomic blast on human beings.[19] The victory made possible by the new and unfamiliar weapon changed America's cultural consciousness about atomic power: immediate and ominous military necessity eclipsed prewar science fiction images of the atom as a benign fountain of limitless energy.[20] To this day, no image of war is as widely recognized, as powerful, or as chilling as the mushroom cloud of an atomic detonation. It became the icon of the new era.

Scientists played a crucial role in molding national opinion as the nation recovered from its shocking wartime introduction to the atom. The public disclosure of the role that physicists had played in the Manhattan Project elevated them to the status of "high priests of a state religion that promised social progress by means of made-to-order technological advances."[21] David E. Lilienthal, the first chairman of the Atomic Energy Commission, wrote that the atomic scientists were modern heroes in an age when science and technology promised to solve intractable social problems.

> The Atom became all important, and so therefore did ... the atomic scientists. These men ... were suddenly catapulted into the very center of human affairs.... The scientist seemed to take on some of the attributes of his world-shaking creation; there was, in the public mind, something unearthly, something superhuman, something uncanny about him.[22]

These atomic technocrats offered sanguine forecasts about the future of atomic energy, while their aura of infallibility gave credibility to their forecasts about the social and scientific implications of atomic energy. A few scientists were uncomfortable with their newly exalted position, but far more embraced their role as ambassadors for the atom. These scientists became for the public the visual, comprehensible, human symbol of a technology that remained essentially inaccessible and unimaginable for most Americans.

The future, remarked *Newsweek* magazine, "would make the comic-strip prophecies of Buck Rogers look obsolete." Experts from the Electronic Corporation of America and General Electric predicted that one of the first and most important commercial uses of atomic power would be for aircraft and rocket propulsion and suggested that nuclear aircraft would be in widespread service by 1955. The National Advisory Committee for Aeronautics predicted the production of atomic-powered aircraft with eight million times as much energy as conventional jet engines. William Stout, former president of the Society of Automotive Engineers, had high hopes for atomic engines but lamented that he did not believe that atomic energy "[could be] safely harnessed into motor vehicles for at least ten, and possibly twenty years." In haunting similarity to predictions made about radium a generation earlier, medical experts predicted that atomic energy would enable doctors and biologists to cure diseases and even prolong life. Home owners would benefit from their own atomic reactors that would provide both electricity and heat. With the advent of home power plants, people would move out of congested urban areas, and the new atomic era communities would extend for fifty or a hundred miles around a smokeless industrial city center. Hugo Gersback, editor of *Radio Craft* magazine, even predicted that personal atomic generators would cool or warm clothes, depending on the season.[23]

The predictions about the nation's atomic future were inspiring. Although most atomic scientists understood that they were years away from perfecting everyday applications for atomic energy, they reassured Americans that anything was possible. "No predictions," wrote Lilienthal, "seemed too fantastic, whether of the doom of civilization through nuclear holocaust or of a world beneficially transformed through the peaceful use of this great new energy source. Men were convinced that they were living in a world in which only the atom counted, and man was almost incidental."[24] Many atomic scientists, reflecting on the moral consequences of their work, became convinced that "somehow or other the discovery that had produced so terrible a weapon simply *had* to have an important peaceful use."[25] Several guilt-ridden atomic scientist-statesmen, including chief Manhattan Project scientist J. Robert Oppenheimer,[26] tried to distance themselves from their wartime weapons research by promoting peaceful uses for the atom through the cooperation of commerce and government under the umbrella of unbiased, "pure" scientific exploration. At the Westinghouse Centennial Forum in spring 1946, Oppenheimer even claimed that the atomic bomb was really a force for peace because society's fear of the bomb would force it to embrace atomic energy's positive attributes.[27] These scientists were a core of influential experts around whom the atomic energy supporters rallied to promote the new technology. Moreover, the postwar progress package that defined the contours of atomic development in the early years remained the ideological foundation for the civilian nuclear power industry.

Faith in the fundamental benefit of atomic power was not the only influence on AEC behavior. Turning the atomic dream into a real, viable civilian nuclear power industry proved more difficult than making optimistic forecasts. The 1946 act had been "written in a rare moment of selflessness," but within weeks after its passage, the government's civilian atomic idealism was already sinking under the weight of formidable obstacles to atomic power.[28] Despite the act's emphasis on private development, domestic atomic energy and the fabulous new world it promised would be possible only with aggressive federal action to regulate and promote nuclear power.

Although the postwar public favored development of peaceful uses of the atom, private atomic power remained a secondary priority for atomic planners during the 1940s. The newly appointed AEC commissioners were confronted with several, more pressing problems. The wartime scientific personnel who had developed the atomic bombs were leaving. This drain of scientific talent jeopardized further research and development. The commissioners also faced a management nightmare. The AEC, noted Leslie Groves, the dynamic army commander in charge of the Manhattan Project, "owned more real estate, plants, and equipment than General Motors; ran more buses than the City of Philadelphia; and had land holdings larger than Rhode Island."[29] Its facilities stretched from Tennessee to Washington State. There were roughly 38,000 employees and private contractors involved in research and manufacturing. The practical challenge of developing a management structure that could efficiently run such a vast enterprise was daunting.

Raw material problems further frustrated civilian atomic power. The commissioners learned soon after passage of the act that America suffered from a shortage of uranium ore and thus a shortage of fissionable uranium isotope. With all available atomic material directed toward weapons production, civilian atomic power was likely to be extremely, perhaps prohibitively, expensive. Moreover, experts predicted that the production of a new generation of "breeder" atomic reactors, which could make civilian power profitable by manufacturing fissionable plutonium fuel, would take a decade of determined research to develop and at least a few more decades to accumulate sufficient fuel.[30]

The pattern of government contracting discouraged private power development. Manhattan Project administrators had believed that "contracting with a few of the nation's largest and best qualified companies and universities was the most expeditious and effective way to develop, design, and produce atomic bombs."[31] That wartime model dominated the AEC's approach to private development after the war, and the government continued to contract with a handful of companies to develop primarily military, rather than civilian, applications. Those select companies also showed little enthusiasm for pursuing peaceful atomic power. Despite the optimism about civilian applications of atomic energy, politicians and utility executives avoided complicated and expensive economics of domestic atomic power generation, and plans for civilian programs remained subordinated to military needs.[32]

Secrecy hindered release of classified atomic technology to private industry. Assessing the act two years after its passage, the critic Byron Miller noted that the secrecy surrounding atomic matters had "not only resulted in great timidity in declassification to the detriment to research, but it ha[d] enabled the Commission to follow undemocratic practices in so-called 'loyalty procedures,' thus driving away many young and able scientists." Miller also noted that the act had shown "schizophrenic performance, with a definite swing to military emphasis despite the victory for 'civilian control.'"[33]

In the months following passage of the act, it was clear that the promise of civilian, democratically controlled nuclear power would go unfulfilled in the immediate future. Weapons fabrication and the procurement and production of nuclear materials consumed most of the AEC expenditures.[34] Some scientists and public officials even feared that nonmilitary applications had been "oversold" to the public. Although Oppenheimer was upbeat about the long-term prospect of nuclear power, in 1947 he appealed to his fellow scientists to temper public expectations by explaining the difficult problems confronting the AEC's peaceful atomic program. Citing military needs, material shortages, and technological hurdles, California Institute of Technology president Lee DuBridge recommended issuing a statement to inform the public that commercial atomic power was years away. Fermi was even less optimistic about nuclear power. He believed that international tensions would drive the nation's atomic program toward weapons production. Under the circumstances, he suggested, it would take fifty years for nuclear power production to meet public demand. Lilienthal feared that the Commission's credibility would suffer when the public learned the truth about the daunting obstacles facing civilian atomic power. He worried that the backlash rooted in public disappointment would bolster the position of those who remained committed to military control of the nation's atomic program and "finish off the rather fragile life of civilian direction of this project." He reflected privately that "deflating the atomic power overoptimism would be definitely in the interest of the Commission."[35]

In late 1949 world events caused AEC officials to reexamine their preoccupation with military applications and weapons stockpiles and adopt a more favorable outlook on civilian power. Beginning on September 3, 1949, American long-range reconnaissance airplanes monitoring the atmosphere over the Pacific reported higher than usual traces of radioactivity. The only possible conclusion was that the Soviet Union had exploded its own atomic bomb.[36] America's atomic monopoly, the military foundation on which postwar planners sought to maintain world order and halt the spread of communism, had ended. In December American confidence was shaken again when Chinese communist forces led by Mao Zedong crushed Chiang Kai-shek's Nationalist army and occupied Beijing, signaling the victory of communism in one of the world's most populous countries. Less than a year later, North Korea attacked South Korea, plunging the United States into a war against communist expansion in Asia.

In response to these incidents, the AEC reassessed the nation's domestic atomic program. It undertook a comprehensive program to expand its acquisition of atomic raw materials for weapons production. These national security emergencies also led AEC officials to view commercial atomic power as an essential component of America's national defense. In the crisis atmosphere, private reactor development appeared more appealing to government and industry planners. The heightened Korean war production demands for electricity demonstrated the possibility of increasing consumer demand for electricity in the near future. A government-sponsored uranium procurement program eased the raw material shortages. Momentum built within Congress and the AEC to pursue a domestic nuclear power program. So complete was the AEC's new commitment to private development that, as early as 1951, the legal scholar James Newman noted, "[T]he preference shown by Congress for the maximum safe participation of private industry in atomic matters has been interpreted by the AEC as dogma. . . . The Commission has elevated private participation above every other principle except possibly security."[37]

National pride also helped to persuade atomic policy makers to support private atomic energy programs when, in late 1952, they learned that Great Britain and the Soviet Union were pursuing their own research into atomic power. The election of President Dwight D. Eisenhower and a Republican congressional majority in fall 1952 solidified the growing trend toward privatization of the atom. The new president favored commercial nuclear power and publicly announced his administration's support for the development of civilian power in his "Atoms for Peace" speech delivered to the General Assembly of the United Nations in

December 1953. Finally, congressional attitudes toward commercial power changed after JCAE chairman Brian McMahon died in July 1952. Throughout his tenure, McMahon had remained committed to the government monopoly of atomic energy, and his death, the Republican victory, and the new president's obvious support of commercial atomic power led to a JCAE reorganization that reduced the influence of JCAE members who had been preoccupied with the security implications of sharing atomic secrets with private industry.

AEC officials and the new Republican administration viewed the American power industry, which had been frustrated since the late 1940s by the government monopoly of the atom and disillusioned by the slow pace of private atomic energy development, as a Cold War ally in the fight against communism. With congressional and administration support, in 1953 the AEC officially acknowledged that the government monopoly on atomic energy should end and atomic technology should be permitted to enter the mainstream of private commerce on a licensed basis.[38] Policy makers no longer viewed commercial atomic power as an afterthought of the military weapons program but rather as a cornerstone of the nation's national security.

Atomic energy policy makers became convinced that private atomic energy should be pursued under free market conditions. Although AEC officials reassessed their attitude toward civilian atomic power and embraced private development programs in theory, the 1946 act, as well as the secretive temperament of the AEC, hindered privatization of the atom. Despite the agency's overtures toward atomic power, in March 1952 AEC chairman Gordon Dean told the Economics Club of Detroit that the Commission's priorities were, in order, to increase the stockpile of fissionable material, to build better atomic weapons, to perfect a hydrogen fusion bomb, and to develop atomic reactors for naval propulsion systems.[39] For nearly a decade after the inception of the nation's atomic program, first the military and later the AEC focused their efforts on military applications rather than on developing peaceful atomic energy.

Until 1953 the JCAE had also shown little enthusiasm for civilian atomic power. During the summer of that year, however, it took the lead in promoting domestic nuclear energy and convened an unclassified hearing intended to give Americans a "better understanding of the prospects for atomic power development, along with a fuller appreciation of the problems which must be solved."[40] The hearing was also designed to deflate pie-in-the-sky predictions about the future of atomic power that were so common that even many policy makers assumed they were accurate. During the proceeding, the senators and representatives examined the possibility of balancing national security, private power development, and financial support for private atomic development. Witnesses from the AEC made optimistic long-range projections about atomic power but deliberately downplayed the likelihood that the new technology would provide significant immediate contributions to the nation's energy supply. The tone of the hearings focused primarily on commercial issues, but AEC officials reassured Americans that the Commission could solve public safety issues arising from a domestic energy program.

Although the Commission representatives cautioned against overly optimistic predictions, they also made it clear that nuclear power was a promising alternative to conventional power production. They testified, however, that the success of any program to foster civilian applications of nuclear power would require industry-government cooperation at a level that was impossible to achieve under the terms of the 1946 act. Private utility industry witnesses also favored a government-industry alliance but stressed that industry required assurances regarding fuel and facilities ownership, patent protection, and licensing arrangements if they were expected to invest substantial capital in atomic power schemes. Ultimately, the hearing testimony exposed the shortcomings of the 1946 act and laid the foundation for reforming it to make the nation's atomic energy program more compatible with practical commercial realities. Perhaps most important, the hearing signaled the acceleration of the government's commitment to a new and unprecedented relationship with industry to pursue a private atomic power research and development program.[41]

Promising testimony, fantastic visions, access to classified information, and wishful thinking were not sufficient to make commercial atomic power a reality. The AEC's emphasis on military applications and the atmosphere of secrecy surrounding nuclear technology during the late 1940s and early 1950s had discouraged utility companies from investing in reactors, or even seriously considering them at all. Market forces discouraged investment because the projected costs for nuclear power did not compare favorably with conventional fuels.[42] Electric utility companies that were interested in atomic energy wanted to break the government's atomic monopoly and own and operate atomic generation facilities. As its critics regularly noted, the 1946 act was simply contrary to American capitalistic traditions. Most significant, however, the 1946 act itself was already anachronistic. The monolithic wartime atomic bomb had evolved into a whole family of tactical and strategic weapons, which had been eclipsed by the destructive power of the hydrogen fusion bomb. America's allies and enemies had developed their own nuclear technologies despite the provisions in the 1946 act designed to preserve America's atomic supremacy. The policy of isolation in atomic energy matters threatened the security of the nation and its allies and could result in other nations, perhaps a communist one, taking the lead in developing beneficial uses for atomic power and offering them to the world. After the close of the JCAE hearings in late 1953, there was a growing consensus in Washington and industrial circles that the 1946 act had outlived its usefulness and needed to be rewritten if domestic atomic power was to get off the ground.[43]

With support from the White House, JCAE members, the AEC commissioners, private utility companies, and the scientific community, Congress easily passed the Atomic Energy Act of 1954.[44] The new law retained the basic organizational structure of the 1946 act but otherwise bore little resemblance to its predecessor. The central concept of the 1954 act was that it empowered the AEC to license private applicants to own and operate facilities to produce and use atomic materials, in effect, relinquishing the absolute AEC control over atomic energy in favor of greater private sector participation in atomic development. "Teamwork between government and industry," noted the JCAE when it reported the bill to Congress, "is the key to optimum progress, efficiency, and economy in this area of atomic endeavor."[45] Such "teamwork" meant that the AEC retained control of "special nuclear materials" but was required to make them more available to industry than had been possible under provisions of the 1946 act. On the surface, the new legislation eased the government monopoly of atomic matters, permitted greater international collaboration, and facilitated the exchange of information between the government and industry. Most important, it led to a complex industry-government relationship with the goal of establishing a competitive private nuclear power industry. The vision was clear to the prophets of atomic power: a government-industry partnership with access to political power, capital, and protection would convert America into a nation powered by plutonium-fueled fission reactors. That nuclear infrastructure, they believed, would guarantee the success of American civilization into the next millennium.[46]

The 1954 act changed the legal relationship between the government and industry, but, noted one contemporary observer, it was "a far cry from the world of Adam Smith."[47] Even after the 1954 act eased prohibitions on industrial atomic development, it did so in a tightly regulated and restricted fashion by which the government maintained strict supervision of public and private atomic matters. In all important respects, the federal government remained firmly in control of the nation's atomic energy program.

The 1954 amendment embodied novel ideas about an industry-government relationship, but in reality its provisions were merely a change in degree of a deeply ingrained pattern of cooperation between the private and public sectors. Promotion of atomic energy through favored AEC contractors continued, and, contrary to the act's emphasis on free competition and private enterprise, a handful of companies with military contracts monopolized atomic research and production under conditions that discouraged widespread industrial participation. By the mid-1950s the atomic energy industry and its insurers realized that the consequences of a nuclear accident at a commercial reactor would result in damages far in excess of what they would be able to pay. To help make atomic power more competitive economically, Congress responded with the Price-Anderson Act of 1957, which shielded utilities from "excessive" tort liability that had helped frustrate private development.[48] The AEC retained control of key government research and granted patents only to inventors who could prove that they developed their inventions independently of any relationship with the AEC—a practical impossibility when research and development studies using atomic materials had to be licensed by the AEC. The Commission determined the terms under which private companies would be licensed to build and operate production and reactor facilities. The AEC was empowered to set a "fair price" for nuclear materials, including raw materials, produced by its licensees, and industry could sell its atomic products only to the government or its licensees. The AEC maintained ownership of raw materials and fissionable reactor fuel, the very heart of any reactor program, but was required to lease nuclear materials to licensees at a "reasonable charge."

The greatest problem resulting from the 1954 act, however, was that it solidified the contradictory AEC roles of promoter and regulator first created by the 1946 act. Although the Declaration of Policy of the 1946 act had alluded to private atomic development through free market competition, the 1954 act directed the Commission to both facilitate the development of private atomic energy and regulate it through its licensing oversight power. During the late 1950s, the AEC became the largest entrepreneur in the atomic industry, was the largest consumer of industrial materials and services, promoted the nuclear industry inside and outside the government, and encouraged and subsidized private companies to enter the atomic industry. At the same time, the 1954 act stressed the AEC's regulatory responsibilities to ensure that the nation's atomic experiment did not jeopardize public health and safety.[49] The multiple roles played by the AEC—partner and competitor, overseer and sponsor, customer and subsidizer—created serious conflicts of interest within the agency. Too often, the Commission subordinated its regulatory duties to its promotional activities.

Paradoxically, then, while the 1954 act purported to open new frontiers for the private commercialization of atomic energy, in reality it created a government-industrial alliance in the guise of free market capitalism. The transition from nationalized nuclear power to private enterprise development under the terms of the 1954 act was marked by growing, not shrinking, federal involvement in atomic development programs. The relationship that developed between industry and the government under the 1954 act was not one of equal partnership, but of a government patron and a private supplicant. As so often happens with agencies, the Commission came to identify more closely with the enterprise it is supposed to regulate than with the public whose interests it is entrusted to protect.

In light of the AEC's conflicting position as both promoter and regulator of the nation's atomic experiment, it is perhaps not surprising that the Commission was frequently hostile to its critics. Any agency is naturally reluctant to admit its errors, especially when, as was the case with the AEC, it believes that the future of the nation is at stake. Public review of Commission activities ran contrary to the cherished atomic secrecy at the root of the AEC's power. Close scrutiny might erode public faith in nuclear power, especially if supported by seeming credible evidence. Criticism threatened to damage the agency's reputation as having anticipated and solved nuclear problems. The Commission might find its promotional activities curtailed at the very time they assumed that national conversion to atomic power was of paramount importance. The assumption among atomic insiders that the nuclear energy program was unusual and required uniquely trained managers isolated the AEC and created the impression that the lessons learned from the nation's nuclear program could not be translated into other areas in which the government played a role in technology management, and vice versa. Finally, the AEC was reluctant to admit its shortcomings because of the financial liabilities involved. Indeed, as the mill tailings crisis unfolded, liability was one of the most contentious issues. The AEC failed to anticipate the environmental consequences of long-term contamination or adequately respond to the early warning signs of the impending tailings crisis. Nor did the agency include tailings management and pollution abatement costs in the price it paid for uranium. Consequently, as the cost for the cleanup grew, the agency was even less willing to admit its errors and shoulder its ethical, if not legal, obligation to pay for remedial action.

The AEC's legal and practical dominance in atomic matters was matched by the JCAE, which, around 1955, gradually assumed a greater policy-making role than it had exercised during the preceding years. The JCAE had evolved into both the protector of the AEC and an important force in atomic policy making. According to Senator Clinton P. Anderson, the JCAE was supposed to act "as a board of guarantors, saying to Congress and the American people that all is well, even if many transactions are behind closed doors."[50] In reality, the JCAE adopted the conflicting functions of advocating for the nation's atomic program in Congress and of reassuring the public, and its elected representatives, that the program was in good hands. Rather than operate as a watchdog to prevent AEC excesses, the JCAE more often worked with the AEC to facilitate the nation's atomic energy program. Ironically, unlike most congressional committees, which usually seek to limit the expansion of executive power, by the late 1950s the JCAE favored increasing AEC authority and expenditures for atomic energy; sometimes it even pressured the AEC and the administration to adopt measures that the commissioners and the president did not favor.[51] These conditions, coupled with the climate of secrecy that prevailed within the AEC, meant that atomic development proceeded with little practical participation from the public or their representatives in Congress. Atomic policies frequently did not reflect widespread public support, or even awareness, of the broader issues that underlay those policies.[52]

For a quarter of a century, the AEC ran the nation's atomic program guided by its twin convictions that atomic power was necessary for national survival and that it must be actively regulated and promoted by the government. In the early years of the atomic age Americans were filled with awe and apprehension about their nuclear future. Atomic policy makers believed that they could balance traditional American institutions of free market capitalism and popular sovereignty with new Cold War national defense, scientific advancement, and social welfare objectives. For atomic power advocates, these newer goals not only achieved a status equal to the nation's historical democratic aspirations, but they became indistinguishable. The Atomic Energy Act of 1946 and its 1954 revisions appeared to lay the ideal foundation for far-reaching and sustained government involvement in an industry that was unique in peacetime and which promised to secure the bountiful future made possible by atomic energy.

That assumption proved inaccurate. From the very beginning of the fission era, astute scientists and policy makers recognized they could not possibly anticipate all the obstacles and risks posed by the new technology. The Atomic Energy Acts, therefore, together served two broad purposes. First, like nearly all long-term policy statements, they were predictions about the nation's future with atomic energy and thus defined the conditions under which the nation conceptualized the new technology. The acts seemed to give Americans authority over a force that was essentially uncontrollable, "apparent knowledge about unknowable outcomes."[53] Influential atomic power advocates in the scientific community, private industry, and the government projected onto the nation their vision of how the atomic-powered future should unfold, and their optimistic forecasts about the benefits of atomic energy defined the nation's response to atomic power. The new atomic aristocracy, more knowledgeable than most people about atomic matters, exploiting their status as the harbingers of a new era and operating in an atmosphere of secrecy that prevented widespread public scrutiny of their activities, dictated the way Americans viewed the operation and consequences of the new technology. As James M. Jasper suggests, during the postwar era, atomic energy policy making "became almost a collective fantasy, or at least was driven by fantasy."[54]

Second, the Atomic Energy Acts provided the legal, political, and social means to make those predictions come true. Once the atomic experts successfully asserted their goals for the nation, they used the laws to make atomic energy policy conform to their visions of the future. Their predictions were remarkably optimistic, and even in the face of mounting evidence that atomic energy was unlikely to be commercially competitive with conventional fuels and far more dangerous than anticipated, proponents continued to proclaim the wonderful benefits of the atom. The acts became a form of social control: the public yielded to the atomic experts' discretion in broad policy matters because the experts claimed to know more than the average person; and the experts used that power to plot the course of atomic development for the nation willing to believe in their fantasy. Unfortunately, as in the case of uranium mill tailings, those experts made mistakes and too often did a poor job of directing the nation's atomic program.

Contrary to the dreams of some atomic energy advocates, it was evident even during the 1950s that nuclear power was a dead issue without government assistance. It was more expensive than fossil fuel generating plants, and scientists in the AEC and in the private sector were well aware of the potential health and environmental risks of atomic energy. A successful nuclear power program required substantial federal incentives to industry, including draconian governmental control of the new technology, large-scale research support, sponsorship of uranium exploration and development, and protection of the industry from tort liability. Nevertheless, despite its obvious drawbacks, vocal atomic energy elites in government and industry worked to convince Americans to accept nuclear power as the ideal power source of the future. That group, through the sheer force and persistence of their predictions and the weight of their credibility in atomic matters, fostered an atomic industry for a nation that desperately wanted to find some peaceful, and beneficial, use for the atomic monster it created and which threatened to turn on its democratic masters. Ultimately, this drive to achieve an atomic power industry cost Americans billions of dollars in environmental remedial programs and doomed the nation's atomic enterprise.[55]

In 1948 Robert Niehoff wrote: "The continued effectiveness of the commission's staff and organization is now, and will be in the future, tested in the crucible of intensive internal and external scrutiny. No agency has ever been put such a severe test. It is necessary for the public welfare that the tests be focused exclusively on the achievement of results."[56] Unfortunately, Niehoff's prediction that Commission activity would be heavily scrutinized failed to materialize. On the contrary, the AEC quickly matured into an institutionalized, quasi-independent agency steeped in secrecy. In the Cold War national security culture in which the Commission evolved, the nuclear policy makers understood that they needed to ensure the success of the nation's atomic program. To that end, they controlled the flow of nuclear information to the outside world. The AEC developed a structure and management style that enabled it to exercise power with little public participation or resort to traditional functions of the American political process. The veil that regularly separated AEC conduct from meaningful public inquiry and input created a legacy of distrust that ultimately eroded public faith in the nation's atomic enterprise.

Niehoff was not alone in measuring the success of the nation's atomic development in terms of results only. That single-minded, goal-oriented approach to atomic development interfered with the ability of atomic energy policy makers to fully anticipate and manage problems that arose from America's headlong pursuit of nuclear power. In the end, the coalescence of the conflicting mandates of atomic energy promotion and regulation in the hands of a single executive agency and the JCAE's increasing focus on policy development at the expense of its watchdog responsibilities had profound implications for the future course of private atomic energy development. Public welfare required more than "results" from our atomic program, it demanded long-term foresight about how to balance the benefits of atomic power with a sensible appraisal of its long-term economic, social, and environmental costs. During much of the 1950s, however, America's policy makers, preoccupied as they were with the Cold War and their own visions for an atomic utopia, failed to provide that measured approach to nuclear power.

3

The Uranium Boom

WHEN WORLD WAR II erupted in 1939, nuclear physicists had already accepted the theory that a "slow-neutron" chain reaction of uranium isotope could produce heat to power electrical generators. Some also foresaw that the new atomic force could be harnessed to create an unimaginably powerful weapon, a military consideration that dominated discussions about the future of atomic power during the war. Yet atomic weapons required a rapid and uncontrolled chain reaction to achieve a devastating explosion, a goal that appeared difficult, if not impossible, to achieve during the early war years. One of the most significant barriers to the atomic bomb, in America and abroad, was the acquisition of sufficient supplies of atomic raw materials, first to conduct experiments and then to manufacture the weapons. American estimates in late 1941 of the amount of uranium isotope necessary to sustain a fission explosion ranged from two kilograms to more than one hundred; the eminent physicist Arthur Compton suggested that the critical mass of isotope might even be as great as one or two tons. But because uranium isotope, U-235, accounts for only about 0.7 percent of naturally occurring uranium, U-238, even a modest atomic bomb project would consume tremendous amounts of uranium, and even greater amounts of uranium ore.[1]

Even before Pearl Harbor, American military planners turned their attention to the deserts of the Southwest as likely sources for strategic raw materials. The Metals Reserve Corporation (MRC), a Reconstruction Finance Corporation (RFC) subsidiary, was responsible for acquiring strategic metals. High among their priorities was vanadium, an alloy element used to harden steel. The Defense Plant Corporation, another RFC subsidiary, financed the construction of five vanadium mills between 1941 and 1943. Although few military planners recognized it at the time, the MRC's vanadium experience would prove invaluable to the nation's atomic weapons program. Vanadium was frequently found in conjunction with uranium in ore, and the knowledge the MRC gained in securing vanadium supplies and the milling infrastructure it contributed to the successful Manhattan Project laid the foundation for the nation's postwar uranium boom.[2]

Creating an atomic weapon seemed to many American scientists to be a distant, perhaps unachievable, goal in 1941. They knew that success depended on securing huge quantities of uranium from which they could produce weapons-grade atomic fuel. Germany had its own atomic weapons program, which relied on uranium concentrates from Austria and Congolese uranium ore Germany captured when it conquered Belgium. American popular media reassured the public that "in spite of the enormously greater resources at the disposal of German laboratories, they could not possibly solve the problem [of U-235 isotope production] in less than ten years."[3] American scientists and military planners were far less confident. They feared that German access to raw materials gave the Nazis a six-month to one-year lead in developing their own atomic bomb. The United States, in contrast, had barely enough uranium on hand to initiate its atomic program, much less produce a working atomic bomb prototype.[4]

After the creation of the Manhattan Engineer District (MED) in August 1942, Washington ordered vanadium millers to stockpile the uranium by-product of their vanadium milling operations. Beginning in January 1943, the Vanadium Corporation of America (VCA), under federal contract with the MRC, began extracting uranium from the tailings left over from prewar and wartime vanadium production. The crude sludge produced from the vanadium tailings contained as much as 50 percent uranium, but MED scientists were disappointed by the material and considered it inferior for their weapons research. The Manhattan Project's immediate need for high-grade fissionable material caused Project planners to focus most of their attention instead on acquiring high-quality foreign uranium. General Leslie Groves pursued a deliberate policy of exploiting foreign reserves whenever possible, both because he believed that foreign uranium sources contained a higher uranium content and because he sought to conserve any small reserve of domestic uranium that might exist. Consequently, nearly all of the uranium used to manufacture America's first atomic weapons came from abroad, including twelve hundred tons of high-grade Congolese uranium ore shipped to the United

States by Union Miniere de Haute Katanga from the Belgian Congo in August 1940. That African ore, however, was only a fraction of what was needed for a sustained atomic weapons research and development program in the future.

Manhattan Project officials also considered the possibility of stockpiling domestic uranium from the American Southwest, where uranium-bearing ores had been mined periodically since the turn of the century. The prospects for acquiring significant amounts of domestic uranium appeared dim. Local miners knew that the deposits of high-quality uranium ore had long since been exhausted for radium a generation earlier. They also understood that the low-grade uranium ore was difficult and expensive to mine and process. Nevertheless, the MRC contracted with Union Mines Development Corporation (UMDC), a Union Carbide subsidiary, to survey America's domestic ore reserves. The results were not reassuring. Between 1943 and 1945, UMDC inventoried the known uranium deposits of the Colorado Plateau. Although it reported the discovery of more than one hundred new vanadium-uranium ore locations, UMDC concluded that exploitable domestic reserves of uranium did not exceed 2,500 tons. It was a conservative estimate, but other evidence corroborated UMDC's pessimistic finding. In 1919 the U.S. Bureau of Mines had estimated the reserves to be a mere 900 tons, and Standard Chemical Company, one of America's leading radium producers, had calculated about 2,250 tons, an amount remarkably in accord with that reported by UMDC twenty-five years later. Production of American ore had been stagnant since the collapse of the domestic radium industry in the early 1920s, and low uranium production figures during the interwar period also supported UMDC's gloomy verdict.[5] Even wartime uranium production from the converted vanadium mills confirmed that domestic uranium supplies were negligible. Between 1942 and 1944 America's principal vanadium producers, VCA and United States Vanadium, accelerated their own ore production, bought ore from independent miners, gathered ore from old mine dumps, and reprocessed vanadium reduction mill tailings that had been accumulating at their mills since the 1920s. Together they produced less than 900 tons of uranium oxide and exhausted nearly all their known ore reserves. Out of the 11,590 tons of uranium oxide the government purchased between 1943 and 1947, only about 1,440 tons came from American sources.[6]

The dreary outlook on America's domestic uranium reserves prevailed after the war. Nobel Prize winner Robert A. Millikan echoed a belief widely held by government planners when he told a radio audience that the world's supply of uranium was "easily exhausted" and "should not be used for any major fuel or power purpose."⁷ AEC planners in 1948 predicted that even under ideal circumstances domestic uranium production would not exceed three hundred tons of concentrates annually. James Newman and Byron Miller, two of the chief architects of the 1946 Atomic Energy Act, confided that the nation was woefully ignorant about its uranium resources. In 1948 they wrote:

> [M]ost Americans are not aware that in deposits of source material the United States is only moderately well off. . . . Available estimates of the reserves of uranium in the United States are exceedingly vague. . . . No doubt, more precise inventories have been prepared since 1942 but, . . . this information is not available. Nor do we know the size of our imports, the nature of our supply arrangements, or the scope of our requirements.⁸

Domestic reserves appeared to be inconsequential, so the AEC made little effort before spring 1948 to expand uranium production at home. Moreover, the small quantity of American uranium was expensive to produce. Although by 1948 estimates of domestic uranium reserves were revised upward slightly from the conservative wartime appraisals, the cost of processing the low-grade American ores would run as high as $20.00 per pound, compared to only $3.40 for Congolese ore delivered to the United States. The AEC continued the government's wartime practice of buying uranium abroad and cultivating positive working relationships with overseas raw material suppliers. Under such severe limitations, it was certain that peaceful atomic energy applications would be subordinated to national security needs.⁹

World events underscored the problem of domestic atomic raw materials shortages. International cooperation to limit the spread of nuclear weapons technology appeared unlikely, as did the prospect for cordial relations with the Soviet Union. AEC planners were acutely sensitive to increasingly ominous global events, especially circumstances that threatened supplies of foreign uranium. Most of the uranium for the first atomic bombs had come from foreign sources, but acquiring uranium from abroad required transporting it, a situation that might prove untenable during a crisis. America's $5 billion investment in atomic energy rested on uranium ore production from subarctic Canada and the Belgian Congo, neither of which appeared capable of supplying sufficient raw material for an expanded atomic program. The Canadian mines were so far north that they could be worked only a few months a year. Production from the fabulous Congolese mines was declining and the Belgians were anxious to raise the price of their ore. Under these circumstances, AEC officials reconsidered the advisability of relying so heavily on foreign sources for uranium. During fall 1947, John K. Gustafson, a mining engineer and the AEC's conservative first director of raw materials, became convinced that domestic ore production was essential for the long-term survival of the nation's atomic energy program.[10]

Criticism at home also influenced AEC planners to reconsider their pessimistic attitude toward American uranium. The emerging Cold War made it clear that demand for fissionable material for weapons would continue into the future, and the Colorado Plateau was the most important American source of uranium. Yet western miners were discouraged and angry about what they believed to be the AEC's myopic focus on foreign uranium. These critics of AEC ore procurement policy alleged that the Commission deliberately ignored domestic ore reserves in order to maximize production from the foreign sources. Mining spokesman Blair Burwell claimed that "there is substantial reason to believe that the activities of the Manhattan District were not fully understood, decoded, or utilized by the AEC." The irascible Burwell alleged that when the AEC assumed responsibility for the nation's atomic program, it ignored a secret wartime report that estimated domestic uranium supplies at 11,611 tons, or more than four times the amount forecast in the Union Carbide report. Consequently, he claimed, the AEC had pursued a misguided program of acquiring foreign uranium at the expense of both American miners and national security.[11]

International tensions and strategic demand for uranium, coupled with blunt criticism from miners like Burwell and members of the new Independent Vanadium and Uranium Producers Association, convinced AEC planners to take practical measures to develop domestic uranium reserves. In light of conservative wartime estimates, however, it appeared doubtful that there was sufficient domestic uranium to sustain the nation's atomic weapons program. It seemed even more unlikely that any domestic ore reserves would be mined and processed at all. Despite governmental enthusiasm for private uranium development, the inhospitable Colorado Plateau provided few economic attractions for mining companies. The poor-quality plateau ore was difficult and expensive to process, ore bodies were unpredictable, and there was little infrastructure, such as roads, power, and labor, on which to build the private uranium industry AEC planners envisioned. Mining and milling the low-grade ore were risky financial endeavors demanding large amounts of investment capital, with little likelihood of success under prevailing market conditions. Stiff foreign competition could dramatically undercut the price of American uranium. Finally, the Atomic Energy Act of 1946 dampened the spirit of free enterprise by making it clear that only the federal government would be able to own uranium, raising the possibility that mining and processing companies would make huge financial commitments but be subject to tight government control and forced to sell their output to the AEC. Even *Time* magazine warned its readers that the chances of striking it rich with uranium were small: "Furthermore, if a prospector finds [a good deposit], his troubles have only just begun. The U.S. (like nearly every nation) regards uranium as strictly Government property. At the end of every rainbow, an FBI man is waiting."[12]

Despite these obstacles, AEC planners remained publicly upbeat about the future of domestic uranium. Lilienthal, for example, reassured Americans that the uranium shortage was temporary. In 1948 he told reporters:

> The potential of the atomic age has been sold short several times in recent weeks. Statements have been made and widely publicized, that there is only enough uranium ore to last a relatively brief period and so the prospects of benefit from atomic power must go glimmering. This is simply not so. It appears clear that atomic energy is on a sound basis for an indefinite period of time.[13]

Lilienthal's comments were based more on his faith in AEC exploration and production programs and his conviction that American market initiative would meet the challenge of producing the needed uranium than on hard evidence of extensive uranium ore supplies. Privately he believed that domestic production would never match the quality or quantity of Canadian or Congolese output. The mining industry clearly needed more than promises before it would commit to full-scale uranium production. At the very least, miners required significant and predictable economic incentives to encourage large-scale production of the uranium the generals and AEC officials deemed vital for both national security and domestic nuclear applications.

Congress and the AEC, reflecting traditional American faith in private enterprise, concluded that new uranium reserves could be located and developed best by private industry working for profit. As early as September 1945, Herbert Marks, an assistant to Undersecretary of State Dean Acheson, stressed in a draft of a presidential message to Congress that atomic energy matters ought to be controlled by a central commission that "should be required to adopt policies involving the minimum practical interference with private research and enterprise."[14] The Atomic Energy Act of 1946 did not authorize the Commission to engage directly in mining, and although the AEC was empowered to buy or condemn real property to acquire radioactive source materials, Congress was determined to minimize the AEC's interference with property rights and private prospecting and mining operations. The 1946 act, noted a Senate committee report, did not intend that the Commission "engage in mining operations in competition with private mining activity unless such operations are necessary to insure to the Commission a supply of source materials adequate for carrying out its duties and responsibilities under the provisions of the bill." The report further emphasized "the necessity of encouraging the activities of independent prospectors" and of ensuring that the "traditional rights of and incentives to prospectors are substantially preserved."[15] Within a year after the passage of the 1946 act, AEC commissioners reassured Congress and the public that "it will be Commission policy to purchase ores for its program from private sources and limit direct Government production as far as possible."[16] When it considered the issue of uranium reserves on public lands, Congress indicated that although uranium deposits were reserved to the United States alone, federal claims were subject to all valid private "claims, rights, and privileges" that existed when the act became law. Moreover, the Commission reaffirmed that any uranium deposits the Commission discovered on public land would be made "available for development by private operators on an equitable basis."[17]

Not only was there political opposition to direct government involvement in uranium development, there was also an underlying presumption against the adequacy of a government-run exploration and production program. Robert Nininger, the AEC's deputy assistant director for exploration, captured the government's prevailing attitude toward uranium development:

> The private enterprise of industry, research and educational institutions, and of the individual prospector, geologist, and engineer is essential to the success of any raw material exploration and development program.... [N]o government-operated exploration program, even assuming that hundreds of the most competent geologists and mining engineers were available, could substitute for these private activities.[18]

John Gustafson told the American Mining Congress in September 1948 that the Commission firmly believed that the discovery, development, and production of new uranium ore reserves would be best accomplished by "competitive private industry, under the stimulus of profits."[19] He noted later that the Commission's programs to expand production of domestic uranium were "designed to aid rather than limit the operations of private enterprise in prospecting, ore production, and ore beneficiation [sic]."[20] Jesse C. Johnson, who succeeded Gustafson as AEC director of raw materials, reflected on the conditions that existed in the late 1940s and summarized the problem to the American Institute of Mining and Metallurgical Engineers in 1951:

> [T]he limited experience with government prospecting clearly [points] to the need of enlisting thousands of private prospectors, geologists, and mining engineers who are out looking for all types of mineral deposits. To obtain the type of effort which everywhere has been responsible for mineral development, it is necessary to provide the same incentive—the opportunity for rich returns.... In mining, particularly prospecting and development, the risks are great and the rewards must be commensurate.[21]

To reduce the economic risk and encourage the moribund uranium industry to maximize production from what government planners continued to believe was a negligible reserve of domestic uranium, the AEC on April 11, 1948, announced an ore-buying program designed to stimulate the private location and production of domestic uranium sources. The Commission offered uranium producers several incentives: (1) a ten-year, $3.50 per pound minimum price guarantee for natural or mechanically concentrated ore assaying 10 percent or more by weight uranium oxide; (2) a bonus of $10,000 for the discovery and production of the first twenty tons of high-grade uranium ore containing 20 percent or greater uranium oxide from newly discovered deposits; (3) a three-year, $2.00 per pound minimum price guarantee for low-grade carnotite and roscoelite uranium ores and the government operation of two vanadium-uranium reduction mills on the Colorado Plateau. In June 1948 the AEC also agreed to pay for one year a haulage allowance of $0.06 per ton mile for transporting ore from remote mines to the Commission's buying depots, with a maximum haulage allowance of one hundred miles. Between 1948 and 1955 the Commission undertook its own domestic exploration program and made its findings available to private prospectors. The AEC and the U.S.

Geological Survey agreed to assist prospectors with geologic advice and provide $5 million for exploratory drilling and promised to restart and upgrade reduction mills that had closed after the war. Finally, the Interior Department ruled that, despite the reservation of uranium deposits on public land, private claims would be valid if they contained other minerals in addition to uranium. For its part, the AEC stated that for claims that contained uranium only, it would "take steps to protect the prospector's equity."[22] According to the AEC, its ore development program had "created widespread interest in searching for uranium deposits," and "[m]ining activity in the Colorado Plateau area ha[d] rapidly increased." The AEC also predicted that production would quickly match wartime output, generating as much as three hundred tons of uranium concentrates a year.[23]

That optimism belied the reality of the uranium-buying program. Despite its innovative appearance, the early benefit of the Commission's uranium purchasing program was more psychological than material for western uranium miners. The program reflected the Commission's belief that domestic reserves were negligible, and few government insiders expected more than a handful of domestic deposits would qualify for the bonus. The initiative was intended to encourage discovery of only high-quality ore bodies and production from mines that could produce ore at costs competitive with foreign sources. The AEC program made wholly impractical allowances for the marginal ores that predominated on the Colorado Plateau. The discriminatory $2 per pound price paid for carnotite and roscoelite ores, by far the most common ores on the Colorado Plateau, discouraged prospectors. The program generally benefited large producers and offered little incentive to owners of the small uranium mines that dotted the West. The payments, though offering a predictable return, were far less than domestic production costs at most mines. Outraged Colorado senators Edwin Johnson and Eugene Milliken blasted the AEC for what they described as its "pinch-penny" attitude toward the domestic miners. They demanded better terms from the AEC, as well as Lilienthal's resignation.[24]

The criticism was accurate, but in actuality few, if any, American mines were in a position to challenge the AEC's foreign uranium suppliers; in early 1948 uranium concentrates from American sources cost nearly six times as much as concentrates imported from the Congo. Moreover, the AEC could not raise its buying price for domestic uranium without angering the Belgians and endangering the fire-sale prices it was paying for their African output. The Commissioners assumed that domestic uranium might at best supplement foreign sources in our national atomic program but would never replace them completely. Yet, even with its limitations, the AEC program sent an important message to western miners that there was at least a limited market for their domestic uranium. More important, the ore-buying program set the stage for a wholly government-subsidized mineral rush of a magnitude never seen before or since in the history of the United States.[25]

As the nation entered the 1950s, America's security appeared more precarious than ever. The Cold War nuclear contest between the United States and the Soviet Union intensified. Confronted by the hostile and nuclear-capable USSR, embroiled in a war in Korea that threatened to plunge America into deeper armed conflict with China, and concerned about the future supply of uranium ore from Africa and Canada, AEC planners reversed their tentative attitude toward American uranium. Military preparedness issues dominated their atomic policy making and accelerated the Commission's tentative ore procurement program. The AEC concluded that domestic supplies, however inferior to foreign uranium, must play a dominant role in the nation's atomic program. Several factors contributed to the change. Increasingly, it seemed to atomic policy makers that survival of the United States depended on expanding the production of domestic ore and gradually weaning America's atomic program away from its appetite for foreign uranium. They also reasoned that demand for affordable uranium would be even more acute if commercial atomic power were to become a reality. Personnel changes in the AEC led to a new emphasis on domestic uranium. The expansion-minded Jesse Johnson succeeded the more conservative John Gustafson and injected new vigor into the AEC's tentative domestic uranium acquisition program. More important, Lilienthal, who had long resisted the pressure of western mining interests and congressmen to accelerate the search for domestic uranium, resigned his position as chairman of the AEC in 1950, paving the way for a massive expansion of the uranium purchasing program.

The most important factor contributing to the AEC reassessment, however, was the new appraisal of domestic uranium resources. Despite its shortcomings, the first three years of the AEC domestic uranium acquisition program demonstrated that the wartime estimates of domestic uranium resources were too low. The AEC's 1948 ore-buying program led to new exploration and discovery of previously unknown ore bodies. Many miners also found that their existing claims contained more commercially exploitable uranium ore than they at first thought. AEC planners began to understand that American uranium supplies were far greater than they originally believed and could provide more than an insignificant amount of uranium. They envisioned domestic uranium production levels that might be sufficient to sustain the government's atomic weapons program and, perhaps, fuel a private nuclear power industry, too. On March 1, 1951, the AEC initiated a program designed to extend the terms of the 1948 buying program and encourage even greater development of domestic ores, including the marginal Colorado Plateau ores. Under the new guidelines the AEC guaranteed, through the period ending March 31, 1958 (later extended to 1962 under the AEC's so-called "stretch-out" program), to buy all the uranium ore miners could produce at $3.50 per pound if the uranium content was 0.20 percent or better.

The government's renewed commitment to a domestic uranium procurement program immediately attracted nationwide interest. The uranium boom was on, matching Sutter's Mill, Cripple Creek, and the Klondike as one of the greatest mining rushes in American history. According to one AEC estimate, more man-hours were spent searching for uranium than were spent seeking all other metals since humankind had first begun prospecting. More significantly, the uranium rush was unique in another way; unlike mining frontiers of earlier times, the uranium bonanza was exclusively promoted, supported, and regulated by the government through the AEC. Although the Commission exercised absolute control over the development, possession, and sale of uranium, the uranium rush was marked, noted one historian, by a "gentle paternalism" that encouraged broad public participation.[26] And the public responded enthusiastically. Thousands of people contacted the AEC, the U.S. Geological Survey (USGS), and the Bureau of Mines requesting information about prospecting for uranium. The AEC alone received about seven hundred inquiries each month regarding the program. To meet the demand for information, the AEC and the USGS published a pocket-sized book titled *Prospecting for Uranium* and sold thousands of copies to potential prospectors. Private dealers sold four hundred Geiger counters each month; Sears and Montgomery Wards sold them in their mail-order catalogs. Prospecting activity on the Colorado Plateau, by both small operators and large mining companies, increased dramatically. In 1950 the location of a 20,000-ton ore body made headline news in the western states, but six years later miners had discovered at least twenty-five deposits each in excess of 100,000 tons and a few that contained millions of tons.[27] Charlie Steen's spectacular $100 million Mi Vida strike in July 1952 exploded into the national consciousness, and by 1953 "uranium fever" gripped America. The AEC's vision of a quiet and orderly uranium search became instead a mad, atomic-age carnival in the heart of the Colorado Plateau.

Unlike early mineral rushes, the uranium prospectors did not concentrate in any centralized area. A steady stream of professional and amateur fortune hunters of every kind—white-collar businessmen, desert-hardened Mormon miners, conservative geologists, "busted out" Texas oil wildcatters, grocers, dentists, professors, bachelors and husbands, even a few women and families—loaded their equipment into army surplus Jeeps and headed into the desolate canyons and mesas of the 200,000-square-mile Colorado Plateau, listening for the clicks from their Geiger counters that might reveal the next big uranium strike and a life of easy riches. Pilots armed with the sensitive detection equipment skimmed the plateau's mesas and dashed into the canyons hoping to pick up traces of gamma radiation, a sure sign of uranium. Cowboys and shepherds tucked Geiger counters into their saddlebags—there was even a counter shaped like a pistol for the drugstore cowboys. During the late 1940s, about 200 geologists were active on the plateau; by 1955 there were at least 1,500. In 1953 about 250 people had staked claims, but by 1955 at least a thousand people had done so. Mesa Junior College (now Mesa State College) in Grand Junction, Colorado, conducted a uranium prospecting course. Clothing manufacturers sold outlandish costumes to well-heeled amateur prospectors, including a girl's "diggerette, jr." jumpsuit. Uranium boomtown children played "claim jumping," a rough-and-tumble game in which one child marked out a patch of sand and defended it from other children trying to claim it as their own. The AEC joined the rush with an $80 million prospecting program of its own. By the mid-1950s the uranium rush boiled out of the Colorado Plateau and into Wyoming, California, Nevada, and Idaho. In New Mexico, a former oil con artist, seventy-year-old Stella Dysart, discovered she owned the nation's largest reserve of uranium under her property on the bed of prehistoric Ambrosia Lake. Once-sleepy western towns like Moab, Utah, Grand Junction, Colorado, and Grants, New Mexico, welcomed the atomic industry because it brought jobs and investment to their largely undeveloped areas. The new boomtowns were choked with prospectors, merchants, miners, movie stars, saloon keepers, swindlers, gamblers, rubes, and prostitutes, all of whom wanted a piece of the government's uranium largess.[28]

The most significant difference with early mineral rushes was that the uranium hysteria was exclusively sponsored by the federal government. The AEC paid discovery bonuses—the Grand Junction office paid an average of $200,000 a month in bonuses—and by law was the sole purchaser of mine output, for which it paid a guaranteed minimum price. The AEC spurred the frenzy in indirect ways, too. It instituted the Defense Minerals

Exploration Administration (UMEA) loan program. Miners secured UMEA loans to cover exploration costs of promising locations. If the prospect led to a significant commercial strike, the miner repaid 25 percent of the loan from production profits, while the federal government repaid the remainder. If the prospect was a failure, the AEC canceled the loan altogether. Between 1952 and 1958 the AEC built nearly twelve hundred miles of roads, at a cost of $13.5 million, to transport ore from the mines to the ore-buying stations and mills scattered across the region. Many of these roads eventually became parts of the state highway systems of Utah, New Mexico, Arizona, and Colorado.[29]

Countless Americans who could not personally join the rush rode the uranium whirligig in the stock market, where brokers traded in the hopes and dreams of ordinary people. In the conservative Mormon citadel of Salt Lake City, smooth-talking Jay Walters, Jr., touched off torrid speculation in penny uranium mining stock that went on twenty-four hours per day. The pace was feverish. Fortunes were made, lost, and made over again as the gamblers and brokers traded. Penny stocks skyrocketed 300 or 400 percent in months, days, and even hours. Speculators bought shares over eggs and bacon in boomtown diners, then sold them for profit over sandwiches at lunch counters across the street. Investors bought stock without even knowing where the properties were located for fear of losing out to the next buyer in line who might hit the jackpot. Everybody had hot tips as investment rumor mills went into high gear. So active was the market in uranium shares that on May 24, 1954, more of these uranium stocks were traded in Salt Lake City than the total number of all shares traded the same day on the New York Stock Exchange. In Utah, there had been nine stock registration applications in 1953 but 149 a year later. By the end of June 1954 there were eighty-one uranium firms registered with the Securities Exchange Commission (SEC) or the Utah Securities Commission. The reckless speculation fever spilled over to Denver and to the San Francisco Mining Exchange, whose president, George Flach, reported that the uranium boom brought "the brightest prospects I've seen around here in ten or fifteen years."[30]

Confident predictions from the uranium kings themselves encouraged people with a few spare dollars, or even their life savings, to take flyers in companies like Sun Uranium, King Midas Uranium, Uranium, Inc., Ute Uranium, Lisbon Valley Uranium Company, Big Indian Uranium Corporation, Amuranium Corporation, or dozens of others. Steve Roman, president of Consolidated Denison Mines, asked *Newsweek* readers, "Can you find a better investment than something the government is buying under contract?" When a reporter asked Steen whether fusion technology might undercut the uranium market, he replied in his characteristically robust fashion: "How the hell should I know? I'm still betting on uranium for the long haul." Some of the companies did not even own a shovel, much less valuable uranium properties. Many only owned options to buy properties. But as long as the property was anywhere near a proven uranium mine, investors snapped it up.[31]

A few penny-plungers became rich in uranium stocks, investing during the initial offering and unloading to later buyers, but most did not. The reality was that of the six hundred producing uranium mines on the Colorado Plateau, only about 10 percent were making significant returns. Shady and fraudulent stocks became such a problem that the SEC proposed new regulations in July 1955, designed to curb the worst offenses and ensure that money poured into promotional uranium offerings actually went for the purposes stated in the prospectus and not into the pockets of promoters and company officers.[32]

The successes of people like Steen, Dysart, and a handful of other overnight "uraniumaires" have come to symbolize the great atomic Eldorado. Their dramatic stories of hard work and incredible luck in the inhospitable western deserts reinforced American myths of rugged individualism, pioneering spirit, and "rags-to-riches" success during a time of Cold War stress and uncertainty.[33] Parker-Brothers toy company even included uranium prospecting in a version of its *Careers* board game. But at the zenith of the uranium bonanza, the end was already in sight. Sober and realistic forecasts by scientists, technicians, and mining companies about the technical obstacles to private nuclear power undercut earlier, rosy predictions about the vast demand for uranium.[34] Uranium stocks plummeted in September 1955, erasing millions of dollars in paper values. Worthless uranium stocks ended up at the bottom of safe-deposit boxes, filling wastebaskets, or wallpapering outhouses. The uranium promoters moved on to other schemes. The outspoken *American Mercury* warned its readers of "the coming uranium bust."[35] "The public," said Steen, "has found out what I've known for a long time—that it's a damn hard job to find a good uranium mine. It isn't the bonanza that a lot of promoters led the public to believe. The crooked promoters and brokers killed their own market."[36]

By the mid-1950s the heyday of amateur uranium prospectors and small mining operations was largely over. While Americans reflected on romantic stories of uranium fortunes, large, aggressive mining companies were buying out or forcing out small uranium producers. Big mining companies undertook systematic, professional exploration, and small operators merged into larger, more competitive organizations. "Instead of a headlong scramble for a quick million," reported *Time* magazine, "uranium has grown into a tough, mature business where the survivors are those big enough to find and mine enough high-cost ore to come out ahead."[37] The easy market for buying and selling uranium locations and mines, where a lucky freelance prospector with a marginal uranium claim once could have made a killing, vanished as the established uranium mining companies demanded convincing evidence of proven uranium reserves and predictable economic return before buying a prospector's claim. Venture capital moved away from the volatile speculation of the uranium stock market and into the solid securities of large, established mining firms. In May 1956 Jesse Johnson testified to Congress that ore shipments from the Colorado Plateau would hit 1.5 million tons worth $46.5 million and confidently predicted that output would increase to 2.5 million tons annually. But few small operators shared in the AEC bounty; of the thousand or so uranium mines in the United States, only a handful of men controlled 90 percent or more of the production by 1956.

Start-up costs were especially burdensome and foiled small-time operators. It was not unusual for mining companies to spend up to $1 million acquiring properties with documented reserves before they actually began mining. Homestake Mining Company, for example, once spent $2.4 million to acquire and develop promising uranium properties. Years of feverish prospecting and haphazard claim recording resulted in thousands of faulty or illegal claims, and these imperfect titles often forced mining companies to incur large legal fees to clear their titles to their claims before they could even begin mining. Firms also had to invest money in a variety of other things, including the heavy equipment necessary to wrest the uranium ore from the ground. Easily accessible shallow deposits became harder to find, and the days of the gasoline "mule" engine lifting handfilled ore buckets from small mines disappeared as large-scale mining techniques outpaced the smaller, two- and three-man operations. After the Hecla Mining Company agreed to mine one of the Federal Uranium Corporation's claims, for example, it took $807,000 just to get mining operations started. Utex Corporation's operations at Steen's Mi Vida mine included tunnels that were three stories high and wide enough for eight-ton ore trucks to pass side by side, while Hecla's mine had a three-compartment shaft that lowered miners seven hundred feet underground and lifted thousands of tons of ore a month. In light of the high start-up and production costs, these companies did not undertake mining operations unless they were assured of profits from the AEC uranium purchase program. With the AEC buying-program commitments, financial risk lay not in volatile markets but in incorrectly assessing a particular uranium claim, a mistake that the large mining companies seldom made. Including capital investment, uranium was a $500 million industry in the mid-1950s, its substantial profits derived exclusively from selling uranium to the AEC.[38]

In contrast to spectacular short-term success, long-range predictions about the strength of the uranium market were inconclusive. That uncertainty was compounded by the very success of the domestic acquisition program that had provided a huge uranium stockpile sufficient for the nation's long-term military goals. Consequently, in 1958 Congress refused to continue funding the expensive unlimited uranium purchase program. On November 24, 1958, the AEC announced that after 1962 it would continue to support only those government contracts signed before November 11, 1958. Uranium exploration, by both small prospectors and large mining companies, ended immediately. Without peaceful nuclear applications to replace declining government purchases, the booming uranium industry faced a grim future. Mining companies therefore adopted the view that it made better economic sense to maximize profits under the guaranteed terms of the AEC procurement program than to take things easy and hope for the best in an uncertain future without government price supports. "The passwords in the uranium industry today," commented one mine operator, "are 'big' and 'fast'; you got to be big to have the capital, know-how and properties to be profitable, and you got to be fast because Old Man 1962 [the year the government price guarantee was scheduled to end] is looking over your shoulder." Mining firms that were fortunate enough to have qualifying pre-1958 contracts expanded aggressively. Guaranteed government profits from uranium enabled large companies to afford the capital investment necessary to keep a uranium mining venture in full production. In less than a decade, under the tutelage of the government, uranium mining in the United States had grown from an insignificant industry providing a negligible dribble of radioactive raw material into a vital economic feature of several western states and a pillar of America's Cold War security.[39]

Stimulating and underwriting a private uranium mining industry, however, was only part of the government's uranium procurement policy. Recognizing that without sufficient milling capacity to refine the marginal American ores its domestic uranium program would never succeed, the AEC also created, supported, and regulated a private uranium reduction milling industry to complement its uranium mining program. Like its mining program, the AEC's commitment to the uranium milling industry was unprecedented in scope and set the Commission on a course of governmental action that would have decades-long consequences.

After the low-grade domestic ore was mined, it needed to be processed at a reduction mill to extract the uranium from the matrix of rock that surrounded it. Milling is a simple process by which raw ore is crushed and treated with acid, or less commonly with base chemicals, to leach the uranium out of the ore. Once the uranium is suspended in the leaching agent, additional chemicals are added to the leachate to precipitate the uranium out of the liquid. The remaining solid and semiliquid residues, or "slimes," are disposed in the tailings area as waste. The uranium leachate, which may be recycled through the circuit and re-treated to increase its concentration and purity, is then washed, dried, and packed for shipping in large drums. This semirefined uranium oxide, or yellowcake, is then delivered to reprocessing facilities where the oxide is refined further to obtain uranium metal and fissionable isotope for reactors or weapons.[40]

Milling radioactive ore on the Colorado Plateau dates to the turn of the century, when Poulot and Voilleque constructed their small mill near Summit Creek, Colorado, to extract uranium oxides from carnotite ore for the radium industry. Their American Rare Metal Mining and Manufacturing Company mill produced about fifteen hundred pounds of uranium oxide before it was leased to the Denver-based Dolores Refining Company in 1903 and finally sold to the Western Refining Company in 1904. Several companies, hoping to emulate the Frenchmen's success, constructed their own reduction mills, such as the Dolores Refining Company mill on the Dolores River near Cedar, Colorado, the Copper Prince Mines and Mill Company near Paradox, Colorado, and the Welsh-Lofftus Uranium and Rare Metals Company on the Colorado River at Richardson, Utah. Most of these mills soon succumbed to the high production costs and marketing difficulties associated with extracting uranium from the marginal Colorado Plateau ores, shipping it overseas, and selling it to European buyers. The advent of the booming demand for radium, however, changed the lackluster climate for American radioactive ore. Standard Chemical Company of Pittsburgh, Pennsylvania, beginning in 1912 and continuing through the end of the radium era, operated its Joe Junior carnotite reduction mill on the San Miguel River, a few miles upstream from Naturita, Colorado. At the high point of its production, Standard Chemical owned 375 important claims and employed 200 men in its production facilities.[41]

In addition to radium, Colorado Plateau mines also produced vanadium. The vanadium reduction process closely resembles the uranium milling process. Shortly after the discovery of the rich vanadium-bearing carnotite ores around the turn of the century, several companies, including the Vanadium Alloys Company, the American Vanadium Company, the Primos Chemical Company, the Cashin firm, General Vanadium, and the Electro-Metallurgical Company, invested heavily in vanadium claims. One company, United States Vanadium, constructed a vanadium mill at Newmire (later renamed Vanadium), Colorado, on the San Miguel River west of Telluride. In 1909 United States Vanadium, together with Primos Chemical, expanded the capacity of the Newmire mill, which operated successfully until the plant burned in 1919. During World War I, demand for vanadium for war production created a strong market for the metal and the vanadium companies reaped huge profits.[42]

After the war, demand for American rare metals waned. During the 1920s, vanadium continued to be marginally profitable at best, but slowly the American vanadium companies merged or closed their facilities until only a handful remained. By 1928 only United States Vanadium (renamed U.S. Vanadium Company, or USVC, after its merger with Union Carbide and Carbon Corporation) was regularly producing vanadium. But in the face of such adversity, optimism abounded on the Colorado Plateau. The vanadium companies that had weathered the shake-out of the postwar years saw a bright future for vanadium. In 1927 Rare Metals Corporation of America began work on a vanadium mill in Naturita, Colorado, and operated it intermittently for a few years whenever the cyclical market for vanadium appeared strong. After Rare Metals' merger with vanadium industry powerhouse Vanadium Corporation of America (VCA), the combined companies invested heavily in the future of vanadium, rebuilding the Naturita mill in the mid-1930s and undertaking continuous production at the facility in 1939. By 1940 the mill was processing one hundred tons of vanadium ore per day from its own mines and from independent miners. VCA even built a company town for its mill, Vancoram (derived from Vanadium Corporation of America), near Naturita.

Sensing, perhaps, the first hints of war that darkened the skies of Europe and Asia, USVC officials believed that vanadium would rebound, and on the strength of their conviction bought Standard Chemical Company's antiquated, radium-era Joe Junior mill in 1936 and refurbished it into a state-of-the-art facility capable of handling 240 tons of ore per day. USVC also built its own company town, Uravan, at its mill site. A more disturbing indicator of the upswing in the vanadium industry was the efforts by the established firms to monopolize the vanadium industry. As the outlook for vanadium improved in the late 1930s, independent miners attempted to operate four other small mills in competition with the giants: the Blanding Mines Company mill in southeastern Utah; a small facility near Mesa, Colorado; the Loma mill in the Yellow Cat district of Colorado; and the rebuilt, World War I–era Gateway Alloy plant near Gateway, Colorado. These efforts were doomed by rapacious and barely legal competition from VCA and USVC that crushed these independent refineries' efforts by late 1941.[43] The outbreak of World War II justified vanadium companies' faith in their investments. For the large vanadium companies, mining and milling vanadium for the war effort was a profitable, albeit short-lived, windfall.[44] The wartime production of uranium also proved lucrative for the vanadium firms. Beginning in March 1943, UMDC uranium development efforts included converting vanadium mills into facilities capable of extracting both vanadium and uranium from old vanadium mill tailings. The MRC's VCA-operated mill in Monticello, Utah, VCA's own mill in Naturita, Colorado, MRC's USVC-operated mill in Durango, Colorado, and USVC's own mill in Uravan, Colorado, all had their mechanical and chemical processing circuits modified to extract uranium from their vanadium mill waste. MRC also worked with USVC to operate ore-buying stations to ensure a steady flow of ore to the mills. The Nisley-Wilson reduction mill, a wholly private concern, opened at Gateway in April 1943 to take advantage of the MRC procurement program but was allegedly forced out of business within a year by aggressive USVC market tactics. Manhattan Project planners also contracted with USVC to construct a new pilot mill at Grand Junction to extract uranium by reprocessing so-called green sludge, a semirefined uranium-bearing scum produced by the converted vanadium mills. By the end of the war, these facilities had extracted uranium from three million tons of vanadium mill tailings. Ultimately, nearly two-thirds of the domestic uranium consumed by the Manhattan Project came from the Uravan and Durango mills, 17 percent came from the Naturita mill, and the remainder came from other sources. That amount, however, constituted only about 15 percent of the total uranium the MED used during the war.[45]

In February 1944, when it became evident to strategic planners that the mines and mills of the Colorado Plateau had produced sufficient stores of vanadium for any foreseeable U.S. or Allied war production, the MRC terminated its vanadium contracts and wound up its affairs. Without government support, the vanadium mines and mills first cut back production and then closed altogether. Moreover, the dismal economic predictions for domestic uranium, the government's focus on acquiring inexpensive uranium from foreign sources, and the uncertain future for the nation's atomic weapons and energy programs meant that the newly converted mills would have little uranium processing work in the future. By 1946 the vanadium companies closed all but one of the five vanadium mills, and that one operated at only half capacity. Yet the vanadium firms' wartime experience with the MRC and the Manhattan Project, especially the profits they made from the relationship, made them acutely aware of the potential for government involvement in their industry. If the opportunity arose for another round of government contracts to secure rare metals from the Colorado Plateau, they would be ready.[46]

The vanadium companies did not have to wait long. The postwar boom in uranium mining caused by the AEC's domestic uranium procurement program created a demand for milling services far above wartime production levels. Because it would have taken time and capital to develop and construct new mills, the uranium industry converted existing wartime vanadium mills and the combination vanadium-uranium mills into uranium extraction plants. It was not the best solution from a technical standpoint; mill experts had long known that the converted vanadium plants were inefficient uranium producers. The primary concern of the AEC, however, was for the United States to produce its own uranium as soon as possible, and the conversion of the vanadium mills, despite flaws, was the fastest method of achieving that goal. On May 28, 1947, the AEC entered into a contract with VCA for the delivery of uranium concentrates from its Naturita mill, a contract agreement that lasted until 1958. From that small start, the uranium milling industry blossomed. Beginning in 1951, new and more efficient uranium plants came on line to meet the skyrocketing demand for milling facilities.[47]

Despite the reactivation of the old vanadium mills and the feverish efforts to construct new uranium facilities, milling capacity lagged behind mine production. Two large uranium strikes near the end of 1952 exacerbated the milling logjam, which continued, despite the development of new facilities, for another eighteen months. Carroll Wilson, a former AEC official and, at the time, general manager of Metals & Controls Corporation, noted that "[a]nyone who has been on the Colorado Plateau has seen those piles of ore at the side of the mill, which certainly gives the miner a feeling there is a bottleneck." Wilson later testified before a Senate committee that the AEC had encouraged miners to slow down the removal of ore from the ground because it was so far behind in milling capacity.[48]

The bottleneck was caused, in part, by uranium companies' reluctance to invest in uranium mills. Milling had initially acquired a bad reputation in the uranium industry because of high start-up costs and unpredictable returns despite government guarantees. One uranium industry expert noted that it took about $4 million to build a mill but only about $200,000 to start a mine. Conservative bankers were sometimes reluctant to loan money to finance what appeared to be risky ventures in uranium milling. "Get into mining, but stay out of milling," was the advice of the uranium buccaneers.[49] Many experts in the uranium industry also believed that the industry and the AEC had rushed the mills into production and that, consequently, those new mills had been poorly designed and built without sufficient pilot-plant research. Technological advances improved milling efficiency, however, increasing mill profitability by reducing the necessary per ton investment, regularizing profits, and attracting new investors. These improvements, coupled with the large and steady supplies of ore and government price supports for uranium, spurred a sharp increase in mill construction. Estimated private investment in uranium processing facilities rose from just over $2 million in 1948 to $10 million in 1952 and about $21 million in 1954.[50]

By the mid-1950s many large uranium ore producing companies, like Steen's Utex Corporation, were either planning or had already built colossal mills to process their own ore reserves. The Utex mill in Moab, Utah, the eleventh built in the United States, processed Steen's ore and serviced local mines. When it opened, the state-of-the-art mill was 50 percent larger than any such facility in the region and capable of processing four times more ore than the antiquated VCA mill in Durango could during the height of its operations. Because the AEC was the only customer for the uranium oxide produced at a mill, as a practical matter no mills were constructed unless and until a government purchase contract had been signed. Although the government did not finance the mills and they were privately owned, the favorable tax laws made uranium mills eligible for rapid amortization. Nine mills operated by 1956, including the converted vanadium mills, and five more were near completion. The union between mining and milling that gave America its uranium independence and sealed its atomic world leadership was at last complete. By the time the federal uranium procurement program finally ended in December 1971, the AEC spent nearly $3 billion to buy more than 348 million tons of radioactive material.[51]

In 1942, when the Manhattan Project began, the United States possessed barely enough fissionable raw material to initiate the long and expensive process of designing and building an atomic weapon. Most of it came from foreign sources. The shortage spurred a frantic wartime search for uranium required to make the atom bomb more than mere scientific theory. Lingering misgivings about uranium supplies during the early days of the AEC did not deter the prophets of commercial atomic power, and the government initiated its program to privatize the atom even before it was certain that such an industry would be viable. In pursuing the goal of nuclear energy, the atomic technocrats, from the beginning, "sought to discover the advantages of the technology and to discount its costs in the *absence* of knowledge of its economic or technical practicality. . . . [F]or this technology, social desirability would be decided in advance of performance."[52] This insatiable desire for atomic power in the face of the serious raw material shortages compelled AEC planners to underwrite the largest mineral boom in the nation's history and promote the construction of a series of reduction mills to process the uranium desperately needed to sustain the nation's weapons program and make the dreams of the nuclear power a reality. The nation's uranium acquisition program was wildly successful, and by the late 1950s America was uranium self-sufficient. More significantly, however, it set the stage for one of the earliest critiques of America's atomic energy policy and the acrimonious philosophical battles over nuclear power that would tear through the national consciousness and erode the nation's faith in the atom. The milling of uranium ore generated huge amounts of low-level radioactive waste that contaminated vast expanses of land and water in the United States. That atomic policy makers overlooked or ignored the serious environmental and public health consequences of achieving uranium independence seriously damaged their credibility and undermined the future of the nation's atomic endeavor.

4

Warm Water
Tailings and Water Pollution

IN GRANTS, NEW MEXICO, across the street from the downtown park, a small museum celebrates the region's uranium mining and milling history. Outside the building, mine equipment dwarfs museum visitors. Inside, photographic panoramas, displays, interpretive texts, even a life-size reconstruction of a uranium mine, tell of the uranium boom on the Colorado Plateau as the epic story of humanity's conquest of nature. In the tableau, hard-hatted miners and mill workers are symbols of rugged individualism; mining companies and uranium buccaneers embody the spirit of American enterprise and capitalism; and uranium mills are industrial crucibles pouring forth the uranium oxide yellowcake on which the security and future of the nation depended. Standing astride this vibrant industry like a great colossus was the Atomic Energy Commission, offering wise counsel and benevolent expertise and ensuring that the wheels of the atomic commerce continued to spin smoothly to guarantee the safety of the free world. Not surprisingly, financial contributors to the Grants museum include major mining and milling companies whose legacy is so gloriously retold.

The story of the uranium frenzy that swept the American West is far more complicated than the museum leads visitors to believe. Certainly, it is a tale of epic proportion. The uranium boom of the 1950s was unique among mineral strikes. The stakes for the nation appeared to be very high during the Cold War. The uranium industry succeeded in wresting the radioactive ore from the Colorado Plateau, one of the world's most hostile and unyielding deserts. As is so often the case in history, however, the events and characters of the uranium story were neither so good nor so evil as they first appear.

By the end of the 1960s, America was a place very different from the one portrayed in the Grants museum. It had awakened to the powerful influences of the environmental movement. Authors such as Rachel Carson electrified the public by explaining that our lives depended on the health of our planet. In an atmosphere charged with a new ecological awareness, the environmental shortcomings of the government's atomic program stood out in sharp detail. Yet the nation's rejection of atomic energy was neither immediate nor the consequence only of a national embrace of environmental ideals. Harsh public criticism of atomic energy during the 1960s and 1970s may have sealed the fate of domestic nuclear power, but the seeds of public distrust of atomic energy were sowed years before the public became infatuated with environmentalism.

The failure of atomic energy in America was in large measure the culmination of a long history of growing public distrust of the nation's atomic experiment and those who controlled it. Among the most damning factors for the future of the uranium industry and the fate of atomic power in America was the failure of national atomic policy makers, most notably the AEC, to anticipate and remedy long-term occupational hazards and waste problems created by the uranium industry it founded and regulated. Equally destructive was their attitude toward problems brought to their attention by atomic critics. Although the nation's nuclear elite were technologically skilled and highly motivated to pursue their vision of a fission-powered future, they were surprisingly unable to wage a successful campaign against the growing public disapproval of the atom. One of the early problems that exposed these weaknesses and cast doubt on the nation's atomic program was the growing health and environmental crisis associated with uranium mill tailings.

Atomic energy had a sinister reputation in the post–World War II era because of its military application, but the government and private atomic power advocates reassured the public through a massive education campaign that the risks of the nation's nuclear power program had been addressed and that new technology would solve any future problems. Throughout the 1950s, the AEC was preoccupied, on the civilian front, with reactor safety, accident prevention, and other protection issues relating directly to the generation of electricity by atomic power. In retrospect, American nuclear facilities had a reasonably good record regarding immediately recognizable radiation injuries. The nuclear advocates also downplayed the dangers associated with the new technology and said little about its potential costs in terms of dollars and disease and death caused by exposure to radioactive materials. "[W]hile many of the hazards of radiation were clearly apparent in 1946," noted Elizabeth Rolph, "no one viewed the potential threat of an evolving nuclear technology as intolerable or something good engineering could not acceptably control."[1] The persuasiveness of federal reassurances about the risks of atomic energy, coupled with the fact that most people did not understand the complexities of atomic energy, left Americans unprepared to face the radiation serpents that inhabited their new atomic Eden.[2]

The AEC acknowledged but minimized the radiation hazards in the uranium raw material processing stage. It reported that ionizing radiation hazards in the front end of the uranium fuel cycle could be controlled as completely as any other industrial hazard. One AEC publication noted:

> [T]he amount of radiation met with in feed-material processes is tiny compared to the quantities in later stages of production. . . . Mining and handling uranium ores usually present no special problems, except that adequate ventilation and dust control are even more important than in other mining operations.[3]

However, the agency failed to fully anticipate possible long-term radiation safety concerns in the milling stage of the nuclear fuel cycle. Because of that oversight, thousands of uranium workers and countless numbers of the public were unnecessarily exposed to low-level radiation in concentrations that occupational health experts, atomic scientists, mining engineers, and AEC officials had known for years were potentially hazardous to human health. Moreover, the AEC's persistent focus on short-term, high-level radiation hazards, its favorable interpretation of the scientific evidence regarding the risks posed by low-level radiation exposure, and its self-confidence in its own expertise on all atomic energy matters, including environmental and public health, all worked to divert the attention of agency managers from the explosive public relations disaster arising from mill tailings pollution. Above all, atomic technocrats had a difficult time resolving the tailings issue because they were convinced that they had anticipated and resolved radiation-related health dangers. "The depth of their commitment," noted Herbert Simon, "prevented them from considering objectively whether the evidence was on their side."[4] Yet farsighted wisdom was exactly what was required to manage the long-term hazards of uranium mill waste.

Most of the tailings left the mills as a semiliquid waste, or "slime," generated during the milling process. A common practice in the early days was to dispose of the liquid radioactive mill effluent directly into the rivers and streams after it had been minimally diluted to reduce radiation and acidic or alkaline levels. One mill, for example, regularly discharged tons of solid tailings each day into a tributary of the Colorado River. More commonly, mills stored fluid and slime behind massive earth dikes and dams where the liquids would evaporate, leaving the solids behind. In fact, at many locations mill companies relied on dry tailings left behind after evaporation of the liquid component of the slime to create their mill effluent containment dams. These containment dikes and dams usually had little structural support. They ruptured with "alarming regularity," releasing chemical and radioactive contaminants into nearby watersheds. Most tailings containment locations were unlined, enabling tailings liquids and contaminated water to soak into the subsurface aquifer.[5]

The tailings piles that grew as the moisture evaporated or seeped out of the mill discharge were unstable mini-mountains of sand easily eroded by wind and weather. A few of the piles were located next to rivers. The sandy tailings eroded into the water during the spring runoff, and undercutting from changes in the river's course caused parts of the piles to collapse into the flowing water. Some mill companies actually relied on spring runoff to wash away a portion of their tailings burden.[6] Most tailings piles were in inaccessible areas far from communities, but some were next to towns and cities, including Durango and Grand Junction, Colorado, and Salt Lake City, Utah. Sometimes mill companies sprayed their tailings piles with water to reduce windblown radioactive dust in the air above nearby communities. More commonly, because the tailings piles were constantly growing through normal mill operations, mills simply left them in place and made little, if any, effort to limit erosion into the surrounding environment.

The problem of radiological uranium mill pollution was first identified in the surface waters of the Colorado River Basin. In fact, tailings contamination was the first significant water pollution problem identified in the basin. Radium was the most hazardous of the elements contained in the tailings and was, therefore, of primary concern to public health experts. From the radiological health safety standpoint, tailings were more dangerous than unprocessed uranium ore, because the tailings particles were very small and water more easily dissolved radium out of tailings than out of uranium ore in its natural state. Consequently, the concentration of radium in rivers or streams contaminated by tailings was higher than that normally found in regional watercourses.[7] The radium levels troubled state public health experts.

At the request of several Colorado Basin states, in October 1950 the U.S. Public Health Service (USPHS) evaluated river contamination caused by uranium milling operations. During the brief survey, investigators looked for radium in water samples taken above and below the effluent output of four mills. The results were startling. The soluble radium level in the Animas River, for example, was 21 times higher below the Vanadium Corporation of America mill in Durango than above it, and the radium level in the San Miguel River was 130 times higher below the mill at Naturita than above it. The USPHS conducted a similar survey in September 1955 on rivers near eight uranium mills. In addition to water samples, USPHS investigators collected river bottom materials, algae, and insects above and below the mills and analyzed them for radioactivity and radium contamination. The water samples showed radium concentrations as high as 20 times the levels of background radiation, even though the effluent output from the mills at the time of the 1955 study was lower than usual. The results of the 1955 river water sampling thus were probably an understatement of the actual radioactive pollution problem. The analyses of the biological and streambed samples were even more alarming. Unlike the water samples, which provided a snapshot of water conditions at the time investigators took the sample, the mud and biota samples showed the cumulative river conditions and provided more than a momentary indication of the river's radioactivity. The aquatic insects collected during the 1955 survey had radium accumulation as high as 60 times normal background levels, and the algae samples showed radium content as high as 100 times normal levels.[8]

Two years later the Colorado Department of Public Health, along with the state public health agencies of Arizona, New Mexico, South Dakota, Utah, and Wyoming, conducted a series of uranium mill surveys, in cooperation with the USPHS, to determine the characteristics of the mill effluent and assess radium contamination of Colorado Basin rivers. The 1957 survey revealed that about 0.38 percent of the radium present in the ore was dissolved during the average acid leaching process, of which 0.06 percent remained in the slime and 0.02 percent was found in the finished yellowcake, leaving 0.30 percent of the radium in the liquid effluent. The remaining 99 percent of the undissolved radium occurred in the solid tailings. The statistics for the average alkaline leaching process showed that 2 percent of the radium was found in the yellowcake and 98 percent in the tailings. The USPHS assigned Joe Anderson, an aquatic biologist, to make a bioassay of the Animas River. Anderson discovered that the radioactive and chemical mill pollution was so extensive that below the mill at

Durango there was a total absence of aquatic life. For nearly fifty miles the river was a "biological desert." At the time, as many as thirty thousand people lived along the Animas River below Durango and relied on the river for domestic and agricultural water.[9]

State and federal public health officials were alarmed by the conditions of the Colorado Basin rivers. The pollution also raised difficult questions. Although scientists had known for years about the correlation between radiation and cancer and about the extensive research on radium poisoning and radon gas, several critical issues remained unanswered. Among the most important was whether there was a "threshold" radiation exposure below which people would not suffer long-term harm. Biologists were especially concerned about the potential for harm from the cumulative effects of chronic exposure to low-level radiation. That problem was compounded by practical research obstacles. It was difficult, for example, to adequately measure chronic exposures to individuals or communities over extended periods, especially among the historically transient population of mill laborers. It sometimes took years for cancer symptoms to appear. When people did develop cancer, it was often impossible to conclude whether the disease was triggered by low-level radiation exposure or some other environmental condition, unsafe lifestyle, or genetic defect. As early as 1950, both the National Committee for Radiation Protection (NCRP) and the International Commission on Radiological Protection (ICRP) maintained that there was no safe threshold. Building on that assumption, the AEC attempted to set standards that were expected to prevent harm to most individuals. Yet, despite nearly a decade of research by the AEC's Division of Biology and Medicine and federal contractors, as late as 1962 the threshold issue remained unresolved. These uncertainties caused most radiation safety experts to urge caution in dealing with low-level radiation. In contrast, Commission officials downplayed the dangers.[10]

It was with this watchful attitude that health officials approached the problem of radium water pollution. They looked to the AEC, which had broad regulatory authority over nearly every aspect of America's atomic energy program, to take leadership in limiting radioactive water pollution that they feared might develop into a serious health hazard. The Commission's response was not enthusiastic. Rather than take a forceful and creative approach to evaluate the long-term radium water pollution threat and mitigate the problem, the agency maintained that it was prohibited by law from exercising any authority over radium. The agency's legal experts pointed out that the 1946 act, while charging the AEC with safeguarding the nation from atomic hazards, left development of the specific regulatory details to the AEC. Section 11(x) of the act granted the Commission regulatory authority over "source materials," defined as "uranium, thorium or other material which is deemed by the Commission, with the approval of the President, to be particularly essential to the production of fissionable material." Subsequent AEC regulations interpreted "source material" to be any substance containing uranium or thorium with radioactive content above 0.05 percent by weight. In most ores, those elements combined accounted for only a fraction of the total radioactivity of the ore. Similarly, AEC officials also claimed that the Commission was prohibited, under section 62 of the act, from requiring licenses for quantities of source material that, in the opinion of the Commission, were "unimportant." By regulatory definition, "unimportant materials" were those containing less than 0.05 percent "source materials." Consequently, although the Commission had clear authority over uranium ore before it entered the mill, the agency argued that it was prohibited from regulating tailings as "source materials" because the mill process removed nearly all such materials from the ore and rendered the tailings "unimportant."

The agency argued further that it was not responsible for radium, the most hazardous element in tailings, because the Commission had exempted from federal regulation naturally occurring radioactive materials that were not essential for the production of "special nuclear materials," that is, atomic fuel. Although section 61 of the 1946 act provided the mechanism for including other materials under the definition if they were "essential to the production of special nuclear material," in addition to finding that regulating such materials "is in the interest of the common defense and security," AEC legal advisers maintained that it was doubtful radium was necessary for the production of reactor fuel or weapons, nor did they see any compelling national security interest to redefine the element as source material. Finally, AEC attorneys noted, regulating radium was questionable in view of the specific provisions of the act and its legislative history. In short, the AEC maintained that it had jurisdiction over radiation hazards arising from source material, special nuclear material, and by-product material ("material yielded in or made radioactive by exposure to the radiation incident to the process of producing or utilizing special nuclear material") but that radiation hazards from other sources, including radium in tailings, were the responsibility of other federal or state agencies.

It seemed unlikely that the AEC could lower the 0.05 percent concentration limit and thus make the tailings "source material," subject to Commission jurisdiction. Even the most efficient mills could not remove all the uranium from the ore, and trace amounts remained in the waste stream. Lowering the concentration limit would have required a determination by the Commission that quantities of source material it regarded as "unimportant" were no longer unimportant from the standpoint of public health and safety or the common defense and security. But to take the position that source material in lower concentrations was no longer unimportant, so as to circumvent its lack of jurisdiction over radium, would permit the Commission to extend its jurisdiction to any materials containing even minute quantities of uranium or thorium. Such action would have been questionable in light of the specific provisions of the act, its legislative history, and prevailing AEC attitudes. Moreover, asserting jurisdiction over radium on such a premise could be construed as undermining the explicit mechanism provided by section 61 of the act for including materials other than uranium or thorium under the definition of "source material."

On the basis of such legal arguments, the AEC concluded that regulation of radium and other radioactive naturally occurring materials in the tailings was the responsibility of other federal or state agencies. Yet, because of the historical dominance of the AEC in atomic matters, the climate of secrecy that surrounded the new science, and the unique public health threats posed by huge amounts of low-level radioactive materials, most state health agencies and mineral industry bureaucracies were ill prepared technically or financially to shoulder that responsibility.[11]

The AEC's legal position on tailings was reinforced by the ambiguous atomic regulatory climate of the 1950s that pitted the broad authority of the AEC in nuclear affairs against the traditional role of states in public safety issues. The Atomic Energy Act of 1954 made AEC regulation of private atomic industry necessary and granted the AEC regulatory control over nuclear power and radioactive materials resulting from the fission process. Congress did not define, however, the limits of state participation in atomic matters. The law made the AEC responsible for overseeing most fission-related radioactive waste. By default, individual states were expected to regulate nonfission radioactive materials and devices, such as radium and X-ray machines.

In practice, the distinction between fission and nonfission processes and materials did not provide clear answers to regulatory jurisdiction questions raised by the 1954 act. And the AEC was reluctant to abdicate any of its regulatory duties. States, which historically oversaw public safety concerns, feared that the Commission would exclude them from a role in atomic safety issues. Several states sought to persuade Congress to more clearly define state prerogatives. This standoff between the AEC and states regarding the regulation of fission and nonfission materials made the

AEC even more likely to resist radium regulation for fear of usurping state power. Ironically, though interested in participating in some aspects of atomic regulation, the states most affected by tailings water pollution did not want exclusive regulatory power over tailings. They believed that the Commission was responsible for the mills and insisted that the AEC shoulder a greater regulatory and financial load for a problem the states believed the Commission was instrumental in creating.[12]

Considering the Cold War climate that prevailed during the 1950s, it is perhaps understandable that the AEC declined aggressive tailings management. For most of the decade, one of the Commission's central motives was to ensure a steady, dependable flow of tremendous amounts of uranium to meet urgent national security purposes. Nothing should stand in the way of acquiring atomic materials. Men like Jesse C. Johnson, the AEC's director of raw materials, especially feared that sensational or distorted publicity about health hazards in the front end of the uranium fuel cycle might disrupt the flow of raw materials at home or abroad. Commission managers made a simple risk assessment, weighing the benefits of determined public health action on front-end safety issues against the possibility that those actions might interfere with the flow of raw materials. Activities such as aggressive management of mill waste problems must not occur if they might interfere with uranium production and thus compromise the nation's atomic enterprise. Commission interpretations of the Atomic Energy Acts and its own regulations merely afforded the agency a legal foundation for what it was inclined to do anyway. Only the most compelling safety issues warranted the AEC's aggressive action. Managing tailings for the possible long-term health hazards they posed to mill workers and the public was a low Commission priority.[13]

Given the AEC's virtual preemption of nuclear regulation, the vast practical power it exercised in the nation's atomic program, and its clear regulatory jurisdiction over raw materials in the uranium-ore processing facilities, it had at least two methods to force millers to curtail their pollution—if environmental protection from tailings had been an AEC objective. The first was through its contracts with uranium millers, an approach the AEC never seriously considered to limit tailings water pollution. The AEC maintained that the contracts contained no environmental quality provisions and that it therefore lacked contractual authority to force millers to control their tailings. The AEC also implied that it could not modify the procurement contracts to incorporate environmental protection safeguards.[14] AEC general counsel Joseph Hennessey summarized the Commission's position on mill contracts:

> The tailings piles . . . are the result of the operations of private industrial concerns which have been engaged in the milling of uranium concentrate for delivery to the AEC under fixed-price procurement contracts. These contracts were a private risk venture involving the operation of privately-owned mills on private property for the purpose of milling privately-owned ores. The Government's only business interest was as purchaser of the product. Under the terms of these contracts, the Government has no contractual right to direct or control the manner in which the mill operator disposes of his mill tailings . . . [or] impose new requirements upon a contractor's performance where no right to do so had been reserved in the contract.[15]

The Commission's position was disingenuous. The AEC stressed the private nature of its contractual relationship with uranium millers to justify its tentative approach to tailings management, but the industry was hardly an example of free market capitalism. The mills were privately owned and operated, but even in the most favorable light they were quasi-public operations. They enjoyed the unprecedented patronage and political protection of the powerful AEC. The "risks" taken by the uranium industry were negligible. Ore supply was never an issue for many "captive mills" that were built primarily to process ore from a single mining company. The Utex mill near Moab, for example, usually processed ore from Steen's Mi Vida mine, while the Rio De Oro Company mines supplied more than 99 percent of the ore processed at the Homestake–New Mexico Partners mill near Grants.[16] A company could build a reduction mill with little financial peril because it would not invest in a mill until it had received a government contract, and once the AEC finalized the contract it granted the necessary licenses almost automatically. Since processing uranium from the low-grade ores found on the Colorado Plateau required an enormous amount of capital but was financially precarious without price supports and protection from foreign competition, the government induced private industry to undertake the task by guaranteeing its profitability. The AEC further reduced the financial risks to uranium industry profits by tolerating lax environmental standards. The Commission, which at the time employed the world's leading experts on atomic matters, permitted mills to dump their radioactive tailings with little restriction and without adequately considering that such dumping might create long-term health and ecological hazards in the future. The ability to dispose of hazardous mill waste with minimum regulation reduced uranium processing costs and boosted profits.

It is true that the AEC's "only business interest" with respect to the uranium mills was as a purchaser of uranium concentrates, but the singular business of the uranium milling industry was to produce yellowcake for the AEC, which, by law, was the only entity able to possess "source material." Given its undisputed power in the atomic energy field, its ability to act without effective public scrutiny in atomic matters, and its position as the sole purchaser of yellowcake, it is difficult to believe that the AEC could not have found the means to persuade uranium millers to agree to contract changes requiring improved waste management practices if it had wanted to do so.[17]

The other avenue for controlling tailings pollution resulting from the operation of uranium mills, and the one ultimately favored by the AEC, was through the Commission's licensing power. As a practical matter, the AEC issued mill licenses to anyone who secured a contract with it to provide uranium, and throughout most of the 1950s the AEC did not take into account in the licensing process radiological safety considerations regarding the release and storage of mill effluent. By 1957, however, the evidence on mill-generated water pollution was clear. That radium and chemical contaminants threatened domestic and agricultural water supplies in the arid Southwest, where access to clean river water was the difference between commercial success and failure, even life and death, accentuated the public concern about mill pollution. Cautious state health departments pressed the AEC to take decisive action to regulate mills.

Although the AEC was slow to respond to what it believed was a low-priority radiological health threat, in February 1957, nearly seven years after the first information about radium contamination in the Colorado Basin rivers surfaced, the AEC issued 10 CFR Part 20. That regulation, designed to address a broad range of safety and health issues arising from the more liberal access to atomic technology outlined in the 1954 act, amended the agency's historical approach to radiation safety, including its mill licensing practices. It authorized the agency to use its licensing power to limit the radiological contamination produced by its licensees and addressed control of mill effluent released into "unrestricted areas," that is, into areas that were outside the mill property. The regulation required uranium millers to comply with AEC radiological safety standards as a condition of issuance or renewal of their licenses.[18] For the AEC, it was an ideal solution to a thorny issue. The agency focused attention on the future conduct of its licensees, avoiding responsibility for remedying past problems. The solution also avoided any costly action by either the government or the millers. Ironically, even though many state officials believed the AEC was years late in taking effective action, the AEC used its new licensing policy to enhance its reputation as conscientious manager of the nation's atomic program by claiming credit for solving a difficult environmental problem. For the downstream water users, the new license standards promised to eliminate the worst mill pollution.

The AEC's new position seemed to confirm the states' sentiment that it could have modified its regulations sooner if it had been interested in mill tailings. Before 1957, however, it may have been too much for states to expect the agency to operate at odds with its basic priorities. The new Commission position reflected a change in the agency's fundamental regulation outlook that may not have been possible earlier. Because the 1946 act prohibited private industrial development of nuclear energy, the agency set safety regulations only for its own contractors. For the next several years, the Commission operated according to what the historian Howard Ball refers to as a narrow "culture of safety" that focused its attention primarily on the "safety of AEC labs doing research, development, and testing of . . . risky nuclear technology." To consider safety issues outside that context was "considered unwise by policymakers committed to the development and maintenance of a strong nuclear deterrent."[19]

The agency's "culture of safety" slowly changed, however, beginning in the mid-1950s. The 1954 act broadened AEC responsibilities by making atomic technology available to industry and in so doing expanded AEC regulatory power to include private atomic programs. After the law went into effect, the AEC reexamined its regulatory standards and began the process of designing new regulations to shield private employees and the public from radiation threats associated with nuclear power. Uranium mill tailings control, while of paramount importance for many state health experts, was one of many new protection issues confronting the agency. Moreover, the success of the domestic uranium procurement policies led to a subtle change in safety attitudes toward mills. By the late 1950s the Commission was no longer concerned about the adequacy of America's uranium supplies. In fact, the nation had so much uranium by 1957 that the AEC announced, much to the distress of the uranium industry, that it intended to wind down its uranium purchase program. Under the circumstances, AEC managers undoubtedly felt that closer attention to mill regulation was no longer outweighed by potential supply disruptions.

Although the AEC determined that the appropriate means to control pollution from operating uranium mills was through its licensing power, it did not vigorously enforce its new environmental quality standards at the mills. By the late 1950s the AEC's atomic promotional philosophy caused it to identify more with the milling industry it fostered than with the public it was supposed to protect. Having taken official action to modify its licensing practices, the AEC remained reluctant to take aggressive steps to regulate mills with which it was so closely allied, especially when Commission experts believed that the radiation threat from tailings was low. State health agencies believed decisive AEC action was long overdue and pleaded with the Commission to energetically enforce Part 20's environmental standards for mill licensees. They were especially concerned about the VCA mill in Durango. That mill had been dumping radioactive waste into the Animas River, causing acute interstate water pollution. Radium levels were "disturbingly high" in the municipal water supplies of Cedar Hill, Aztec, and Farmington, New Mexico. Because radium tends to concentrate in living tissue, especially bones, health officials were also concerned about food crops irrigated by the contaminated river and domestic animals drinking the water, the consumption of which could lead to higher radium concentrations in humans living in the region.[20]

Frustrated by what it saw as AEC foot-dragging and the slow pace of cleanup efforts, the New Mexico Department of Public Health took matters into its own hands in April 1958. Pursuant to the Federal Water Pollution Control Act of 1956, it requested the USPHS to conduct an enforcement conference to consider abatement of radioactive contamination of the Animas River, one of the most seriously polluted of the radioactive rivers.

Commission representatives at the Animas River Conference reported that the agency was diligently using its mill licensing procedures to enforce Part 20's regulations limiting release of radioactive materials into unrestricted areas. They pointed out that in February 1957 the Commission began inspecting mills for compliance with the new Part 20 standards. After investigating twelve of the twenty-five operating mills, AEC examiners discovered that "[n]o mill [was] in compliance with the regulations." They found that radium levels in rivers contaminated by mill operations exceeded the federal maximum permissible concentration of $4(10-9)$ picocuries per milliliter. Only the mill at Uravan, however, seriously exceeded that radium concentration level three miles downstream from the plant, although investigators acknowledged that the values were tentative because river conditions varied widely.[21] Yet the findings were encouraging to Commission bureaucrats because they confirmed the agency's fundamental position. The AEC concluded that radiation exposures, although too high, were not "an immediate hazard to the health and safety of the employees [or] to the public." It also tried to deflect criticism and implied that millers were partially responsible for any compliance disputes. Mill regulations, the agency said, "place the responsibility on the licensee to determine the various limits [for] effluent released into unrestricted areas." Further: "We [the AEC] have cited them for failing to determine the concentration of radioactive material in the effluent released in the unrestricted areas." In several cases the agency charged that mills themselves had not been sufficiently diligent in complying with the regulations.[22] However, close questioning of the AEC officials confirmed the states' opinion that the Commission had been slow to compel uranium mill compliance. Further inquiries revealed that the AEC had not issued any citations but was merely in the process of investigating the violations and discussing the matter with uranium millers.[23]

The Commission was confident that it was acting as quickly as possible and viewed the states' hostility as unreasonable. The new standards of Part 20 needed to be explained to mill companies; between 1957 and 1960, the AEC held four meetings with millers to discuss the new regulations and provide training to mill personnel in radiochemical analysis techniques. As provided by its "Rules of Practice," the AEC notified mills of violations and provided a generous timetable for them to remedy deficiencies. Follow-up inspections needed to be scheduled. AEC compliance orders allowed for an appeals process for mills that disputed its findings. Uranium milling was a heavy industrial process involving complicated mechanical and chemical operations. Mills had to uphold their existing contracts. They consumed mountains of raw materials. They produced a semiliquid waste that was difficult to contain. Under the circumstances, new techniques for tailings management would take some time to implement. The Commission believed that any apparent lack of diligence on its part was simply the inherent lag between issuing new regulations and the reasonable time necessary for millers to change their production and waste disposal habits. In any case, the AEC believed that although aquatic radiation levels were high, tailings water pollution was not an immediate health threat. The agency saw little need to expedite the compliance process for what it continued to believe was a low-priority problem.[24]

State and federal public health officials viewed matters differently. The AEC's assertion that tailings were not an "immediate hazard" fueled mistrust. Competent scientific authorities generally agreed that there was no substantial radiological risk associated with the tailings in the short term. For public health experts, however, "immediate" threats from tailings pollution were not the real issue; they already understood that "immediate" hazards were a minor problem compared with the possible consequences of long-term exposure to low-level radiation. They interpreted the AEC's focus on "immediate hazard" language as an attempt to avoid dealing with potentially serious long-term radiation dangers. They remained convinced that the AEC would not assume leadership in developing a comprehensive, long-term solution to the tailings pollution problem. Rather than act with the speed that the states demanded, the AEC only required mills "to initiate comprehensive programs of environmental sampling and radiochemical analysis to identify problem areas." Given that the AEC did not consider tailings to be a serious hazard, the Commission allowed "time for the mills to achieve compliance in an orderly manner so long as mills continue to demonstrate adequate progress toward compliance."25 Robert Barker of the AEC Division of Licensing and Regulation summarized the AEC's close relationship with the uranium milling industry, the basis of AEC response to tailings pollution.

> [O]ur AEC Operations are not of the clobbering type of operation, they are of the cooperative program type. We do have regulations which must be met. To do this . . . we will cooperate with the mills in achieving compliance; we will not, in a sense, move in with the idea of trying to put them out of business because that would in turn put us out of business.26

AEC officials underestimated the negative consequences of their testimony at the Animas River Conference. In marked contrast to the AEC's mill-friendly approach to pollution control in the Colorado River Basin, federal and state public health agencies continued to insist that it act quickly to enforce mill compliance with regulations. The USPHS was particularly determined that radioactive waste pollution abatement be pursued vigorously. Moreover, rather than reassure regional water users about the agency's commitment to clean water, the conference testimony instead underscored for observers the AEC's apparent lack of interest in pollution abatement and proved to be embarrassing for agency officials. While the conference brought to light several important issues and clarified details of AEC enforcement practices, it did little to accelerate the pace of AEC action on mill regulation. To the frustration of state health authorities, the agency continued to follow its deliberate, cooperative procedures to bring mills into compliance with the safety regulations. Under a cloud of state enmity, the AEC in 1959 ordered uranium mills to "assure that concentrations of radioactive material in mill areas and in wastes discharged into streams are brought within permissible limits."27

Only then did the mills take meaningful action to reduce the tailings pollution levels in Colorado Basin rivers that at least stopped the absolute degradation of the ecology of the Animas and other rivers. In spring 1960 the AEC general counsel finally sidestepped the contentious radium jurisdiction issue altogether. Although the Commission continued to insist that it had no jurisdiction over radium, it concluded that the atomic energy agency had the authority to enforce radiological pollution standards for tailings at operating mills, regardless of the uranium or thorium content of the tailings, because the tailings were an integral part of the licensed milling operations. By the early 1960s abatement programs at operating mills had been largely successful in reducing the worst of the tailings pollution.[28]

No sooner had the Animas River Conference made apparent the problem of radium pollution in rivers than another more troublesome tailings problem arose. When the demand for uranium decreased after the early 1960s and the AEC declined to renew many of its production contracts, nearly half the uranium mills were forced to close. Most companies simply abandoned their tailings piles after they shuttered their mills. Only the facility at Monticello, Utah, had undertaken any comprehensive stabilization when it closed in 1960. Unsecured piles eroded into neighboring ecosystems, especially after the moisture in the tailings drained away or evaporated. Rainwater and snowmelt leached radionuclides and mill chemicals into local aquifers. Wind carried the dry tailings dust far away from the piles. Federal and state public health experts, already concerned about the possibility of harm from long-term exposure to low-level radiation, were worried about the potential health problems from the abandoned tailings piles, primarily those piles in or near towns. They also continued to express concern that dispersion of the tailings onto agricultural land near the mills could lead to elevated radium concentrations in the food chain.[29]

Although the AEC required its licensees to limit tailings pollution from operating mills beginning in 1959, abandoned locations posed a new regulatory problem. The agency governed operating mills through its license power, but once the mills closed, it assumed no responsibility for the tailings piles. If the AEC was going to regulate tailings at abandoned locations, it would need to exercise regulatory control over the tailings themselves; but under existing regulations there was insufficient uranium or thorium in the tailings to warrant direct regulatory oversight by the agency.

It was a difficult question. States insisted that the AEC bear some, if not most, of the burden for managing the increasingly unstable tailings. The

AEC claimed tailings at abandoned uranium mills were outside its jurisdiction. At an internal AEC staff meeting held on October 2, 1963, AEC chairman Glenn Seaborg suggested that the AEC could prevent problems with abandoned tailings in the future by assuming some responsibility to require millers to bury or otherwise control tailings at uranium mills before they closed. Yet that solution would not address the problem of tailings at sites already deserted. In 1965 the AEC reaffirmed its refusal to regulate abandoned tailings. It repeated its claim, based on its review of the available scientific data and the radiation safety standards of the time, that "[uranium] tailings [did] not constitute an unreasonable risk to health and safety of the public and that controls over such tailings for radiological safety purposes [were] unnecessary."[30] The AEC further concluded it was unlikely that individuals would be exposed to radiation above safe limits from the abandoned piles in the future "under any conditions which appear probable."[31] The AEC asserted that no convincing reasons, based on either the health and safety or common defense and security provisions of the Atomic Energy Act, justified an amendment to the AEC regulations that would decrease the 0.05 percent regulatory threshold and thereby give the Commission authority over the abandoned tailings themselves. In short, the AEC reiterated its traditional position regarding tailings and determined that the quantity of uranium in the tailings was so small that it was "unimportant" and that the tailings were thus outside the scope of its responsibility. The irony of the agency's position could not have escaped public health experts: the AEC had acknowledged as early as 1957 that tailings at operating facilities were a sufficient potential health hazard to force millers to stabilize and manage their piles, but those same tailings at abandoned sites a few years later were not threatening enough for the agency to regulate them directly.[32]

The AEC favored control of the abandoned uranium tailings piles but by some other authority, but not because it was convinced that tailings were a radiological health threat. According to the AEC, the dusty and unsightly tailings piles, if located in or near towns, were public nuisances. Commission officials viewed the management of the abandoned tailings as simply "a matter of good housekeeping prudence," but "aesthetics [were] not within [AEC] jurisdiction."[33] The atomic energy agency maintained that the need to dispose of the uranium tailings was no greater than the need to manage wastes from other types of ore processing mills and that uranium tailings could thus be handled in the same manner as any other tailings. The AEC favored cooperative efforts among federal, state, and local governments and industry to devise strategies to limit erosion of the piles, with the states assuming primary responsibility for tailings management programs.

To bolster its conclusions that the tailings were not a significant health threat, in August 1963 the AEC requested the Federal Water Pollution Control Administration (FWPCA) of the U.S. Department of Health, Education and Welfare to draft a report addressing the impact of tailings on the ecology of the Colorado River Basin.[34] The FWPCA designed the study to evaluate the radioactivity of the tailings, predict the possibility of radioactive water pollution caused by tailings, and suggest ways to control the spread of radioactivity from the tailings piles. The FWPCA completed its report, titled "Disposition and Control of Uranium Mill Tailings Piles in the Colorado River Basin," in late 1965. AEC officials felt vindicated when the FWPCA conceded that pollution control efforts undertaken by the operating uranium mills had been effective in reducing radioactive water pollution and that there was, at the time, no significant immediate health hazard associated with uranium milling activities anywhere in the Colorado Basin.[35]

The AEC's relief, however, was far from complete. To the chagrin of the agency, the report reflected the traditional public health concern about the possible, but undetermined, problems of chronic exposure to low-level radiation. Despite favorable evidence about the short-term tailings hazards, the thrust of the FWPCA conclusions was not on immediate health concerns but on the need for long-term tailings control. The report detailed what the FWPCA investigators believed were long-term health concerns posed by the tailings at inactive mills. They were especially worried about radium in tailings, because the element was long-lived, leached and dissolved into surface water, and was chemically similar to calcium and, therefore, collected in bones of livestock and people who drank the contaminated water. The FWPCA highlighted the many possible and unpredictable events that could cause the tailings to become a serious environmental hazard in the future. River courses or water uses might change; remote locations of some tailings piles might become inhabited because of population increases or new land use practices; and background radiation levels might rise naturally or because of nuclear fallout, making even small additional doses of tailings-generated radiation exposure more hazardous. The FWPCA urged care, noting that scientific studies and historical experience showed that radiation protection standards were continuously revised, usually becoming increasingly restrictive as additional knowledge about biological effects of radiation exposure came to light.[36]

The FWPCA concluded that its report made evident "a need for caution and a conservative approach to the problem of disposition and control of [inactive] uranium mill tailings."[37] It recommended that interim measures be undertaken immediately to prevent erosion and spread of uranium mill tailings and that distribution of mill tailings off the mill property for any reason be halted until adequate precautionary procedures could be established. Interim measures, the FWPCA maintained, had to be sufficient to prevent the erosion of the tailings for between ten and twenty years, the time the FWPCA assumed would be necessary to investigate future trends, river hydrology, safety standards, and governmental and private responsibility and to establish long-term tailings management strategies. FWPCA officials estimated that the cost of the recommended interim tailings control would be about $500,000, a small amount compared to the $28.8 million uranium mills received for their 1964 output alone. The FWPCA also advocated that discussions be held and binding agreements be negotiated between industry and state and federal authorities for adequate long-term maintenance of the abandoned tailings piles.[38]

The FWPCA report created a storm of interagency controversy. Its recommendations so alarmed AEC officials that they tried indirectly to persuade the FWPCA not to release the document publicly. They suggested that the report remain an internal "Discussion Paper" rather than a public statement of the official FWPCA position regarding the tailings. Although the AEC claimed it wanted to reconcile the differences in opinion between it and the FWPCA, the Commission was more interested in convincing the FWPCA of what it believed were scientific inaccuracies and unwarranted conclusions contained in the report. In a point-by-point analysis and refutation of the report, the AEC repeated its claims that radium in the tailings did not pose an unreasonable health hazard and argued that, from a radiological safety standpoint, the report did not adequately prove the need for interim, much less long-term, tailings management. The AEC suggested that the FWPCA address its objections and revise the report. The FWPCA refused. Neither side was willing to compromise its position, and the agencies were at an impasse.[39]

During this internal controversy, the Department of Health, Education and Welfare held a conference in Grand Junction on December 13, 1965, to discuss tailings pollution problems in the Colorado Basin once again and invited the AEC to attend. The AEC braced itself for conflict with the FWPCA over the contents of its report and, before the conference, hoped to smooth over their differences about the need for long-term tailings management. AEC officials were not hopeful that a common ground could be reached and conceded that "[o]n the basis of past experience with . . . this problem, we do not expect the matter will be fully resolved, but we feel the effort should be made."⁴⁰ At a private meeting before the conference, FWPCA and AEC representatives agreed to avoid airing their differences of opinion during the conference. The AEC representatives also requested that the results of the FWPCA report be kept confidential until the two agencies could reach a compromise. The AEC informed the FWPCA that it was intensifying its program of encouraging uranium millers to work voluntarily on long-term stabilization for their tailings and was providing technical assistance for control programs. Praising this AEC involvement in the management problem, the FWPCA suggested that such voluntary programs would go a long way toward meeting the goals outlined in its report.⁴¹

AEC representatives understood that the December 13 conference would not be public and were therefore surprised to see in the audience federal, state, and local officials together with uranium industry and labor leaders. The conference also received considerable regional media attention. Despite the AEC's request for confidentiality, FWPCA representatives revealed the substantive conclusions of its report. It was the highlight of the meeting and caused a sensation. The Grand Junction *Daily Sentinel* published a balanced report of all the issues discussed at the proceedings, but the account in Denver's *Rocky Mountain News,* one of the most widely circulated papers in the Rocky Mountain region, highlighted the radioactive contamination portions of the FWPCA report.⁴²

The report and the controversy led some officials in the AEC to express doubts about the agency's rigid stance with regard to tailings. A. O. Little, for example, wrote to Seaborg that it was his "personal recommendation ... that it may be appropriate to ask an independent firm to conduct a study of alternative means, along with estimated costs, for disposing of these mill tailings." Little concluded his comments by stating he believed that the AEC's responsibility went beyond legal and technical considerations that apparently left the AEC "in the clear."⁴³ He was unable, however, to convince his peers of the need for an equitable solution to the abandoned tailings problem, and the official AEC response remained doggedly focused on the report's finding that the tailings were a low environmental risk. By stressing the FWPCA's "no immediate hazard" language, the AEC hoped to avoid the divisive and possibly damning aspects of the report and give the impression that the two agencies were in much closer agreement about tailings hazards than they in fact were. Nearly as troubling as the AEC's attempt to sidestep the question of long-term tailings management was the impression the Commission tried to foster that the tailings were a minor menace and would remain so indefinitely. Unlike the AEC, FWPCA officials believed that although the tailings did not present an immediate threat, there were so many unknowns about the harm from tailings radiation that caution dictated a policy of aggressive long-term management.⁴⁴

The impasse between the agencies over the FWPCA report was rooted, ultimately, in neither science nor law but in outlook. Both agencies agreed that the small amounts of radiation emanating from abandoned tailings posed little immediate risk to surrounding communities. The differences lay in their fundamental assumptions about radiation safety. The AEC viewed abandoned tailings as a low-level hazard that did not warrant aggressive federal intervention. Public health officials had a dramatically different idea about tailings. Future uncertainties, the fact that the radiological contaminants in the tailings had long half-lives, and the knowledge that harm from long-term exposure to low-level radiation was not fully understood caused the FWPCA to recommend a conservative, cautious approach to tailings management. As noted in one AEC document, "[T]he [US]PHS recommendations appear to be based on the philosophy of reducing all potential exposures to the minimum practically obtainable and the fact that radium-226 (the radioisotope of major concern) has a very long half life and can be leached from the tailings by water."⁴⁵ Whereas the unpredictability of the future led public health experts to recommend prudence in handling the abandoned tailings, the AEC saw that same unpredictability as justification for avoiding long-term tailings management.⁴⁶

What began as an internal dispute erupted into a major public policy debate over uranium tailings. The largely regional tailings story came to national attention on March 5, 1966, nearly twenty days before the FWPCA officially issued its report to the public, when the *New Republic* ran its article "Uranium Mystery in the Colorado Basin."⁴⁷ The story was sensational, contained several inaccuracies, and implied that radiation pollution conditions in the Colorado Basin were far worse than they were. It overstated the uranium tailings hazards outlined in what the AEC believed was an already greatly exaggerated FWPCA report. Although determining the precise impact of the *New Republic* article is impossible, it clearly focused widespread public attention on the tailings issue. Debates inside federal and state agencies over the hazards of tailings were raging before the article appeared, but after publication, and largely because of it, tailings became the subject of environmental interest and concern on a national scale. The article publicly revealed for the first time the simmering conflict between the AEC and the FWPCA regarding the need for long-term tailings management. It also attacked the AEC, claiming that the agency opposed issuance of the FWPCA report for fear that it would cause "public hysteria," a charge that AEC officials vigorously denied. AEC officials responded that they were the ones who had originally commissioned the study of radiological conditions in Basin rivers and that they had never attempted to suppress the FWPCA report. The AEC quickly drafted a response that highlighted the article's technical inaccuracies and assumed the controversy would blow over.[48]

To the dismay of AEC officials, the negative publicity escalated rather than diminished. Terry Drinkwater, a CBS reporter, aired a lengthy report on uranium tailings on Walter Cronkite's evening news telecast and on CBS radio on April 20, 1966, less than one month after the FWPCA released its report. Drinkwater said dramatically:

> The Atomic Energy Commission which oversees the entire uranium industry says that licensing control of tailings is not required. The AEC does not see a role for itself in doing anything about the tailings. . . . If nothing is done to cover up or remove these tailings and if government health officials are right then another generation may well look at these radioactive man made mountains as monuments to the carelessness of this generation; man's carelessness in the early years of the nuclear age.[49]

Despite inaccuracies, Drinkwater's newscast further emphasized the interagency conflict, with the AEC portrayed as the villain endangering the public health. Drinkwater's report concerned AEC officials and members of the Colorado Department of Public Health, who feared that the negative publicity might disturb local residents and create difficulties for the state agency in developing the state's own tailings management regulations.[50]

The flurry of public interest in the inactive uranium mill tailings problem caught the attention of Democratic senator Edmund S. Muskie of Maine, chairman of the Subcommittee on Air and Water Pollution of the Senate Committee on Public Works. The senator, who had already earned a reputation as a powerful advocate of antipollution laws, held a subcommittee hearing on May 6, 1966, to explore issues of public and private responsibility for radioactive tailings and to determine the extent of the pollution hazard.[51] Muskie also hoped to inform the public of both long-term and short-term problems associated with the tailings. Testimony at the hearing revealed that as of 1966, only one of the tailings piles generated by the thirty-four uranium mills in the western states had been stabilized to limit wind and water erosion.[52]

Testimony by AEC and FWPCA representatives before the Senate subcommittee underscored the gulf between the agencies' positions on the need for long-term stabilization of abandoned uranium mill tailings. Summarizing his agency's position, Murry Stein of the FWPCA testified:

> The question we must face with uranium mill tailings is, whether we assess responsibility for control of the uranium mill piles now while the parties responsible are still readily identifiable. The alternative, it would seem to us, is the bequeathing of a legacy to posterity of tailings piles containing long-lived radioactive materials dotting the river banks of the Colorado River Basin without clearly defined responsibility for their control. . . . It is impossible to predict the changes that may occur over centuries that may affect or alter the importance of the tailings as an environmental contaminant. . . . It is our responsibility and obligation to posterity to control or devise means of ultimate disposal of the radioactive waste materials that we have produced in our time.[53]

The AEC representative, Peter Morris, reiterated the AEC's long-standing position on tailings:

> The evidence available at the present time does not support a conclusion that the uranium tailings represent a radiation hazard to their environment. . . . [W]e find it difficult to conceive of any mechanism whereby the radioactive material which is so widely dispersed, could become so concentrated as to exceed current applicable standards for protection against radiation. . . . We recognize, however, that in their gross physical aspects the uranium piles . . . constitute a nuisance to those communities which are adjacent to them. . . . The Commission plans to continue its cooperative effort with Federal, State, and local authorities and with the milling industry to achieve adequate pollution control.[54]

Under Muskie's close questioning, Morris remained unshakable in his conviction that the AEC was not primarily responsible for the abandoned tailings problem and that the tailings were not an immediate health hazard. After stressing that the tailings did not need to be stabilized for purposes of radiological safety, he contradicted himself by admitting that the AEC could not rule out the possibility that a hazard might develop from the tailings. Morris then contradicted himself again by asserting that agency scientists could not envision any future circumstance under which the tailings could become a radiological health threat.[55]

The AEC came under intense congressional criticism because of the hearing. During the Senate discussion of the hearing testimony, Senator Muskie stated:

> The committee was inclined to accept the judgment of [the FWPCA] that a long-term hazard does exist from erosion of the piles. . . . The Committee feels that the Atomic Energy Commission has not satisfactorily discharged its responsibility toward the prevention of radioactive pollution in the Colorado River Basin. The AEC has a clear obligation to protect the public from radioactive hazards generated by activities it licenses, regardless of the traditional regulation of radium by the States.[56]

The Senate Committee on Public Works, in reporting what became the Federal Water Pollution Control Act Amendments and Clean Rivers Restoration Act of 1966, urged the FWPCA to act expeditiously to establish responsibility for control of the tailings piles. It also requested that the agency assure the committee that any tailings control measures applied would be sufficient to achieve long-term protection of the health of the people who lived within the area of the Colorado River or who depend on it.[57]

The FWPCA report and the Muskie hearing generated a great deal of media attention, most of it critical of the AEC. After the hearing, AEC and FWPCA representatives realized that continued public scrutiny of their differences could damage both agencies' reputations. The FWPCA wanted the issue of responsibility for abandoned tailings settled quickly, fearing that continued negative congressional and public attention would cause the AEC to reduce its already minimal cooperation in the development of a comprehensive tailings management policy. AEC officials worried that additional probing might erode public confidence in the agency and accelerate the increasing public hostility to the agency and the nuclear energy industry as a whole. The two agencies, consequently, sought publicly to reconcile their differences and develop a common position on the need for long-term tailings stabilization and the locus of responsibility for such activities. The AEC, the USPHS, and the FWPCA signed a joint agency position statement on December 8, 1966, purporting to outline their common position on long-term tailings management. The agencies agreed formally that "inactive tailings piles resulting from uranium milling operations should be structurally stabilized and contained to prevent water and wind erosion. Active tailings piles should be managed to minimize such erosion during use." The statement placed the responsibility for containment on individual mill owners, who were to develop specific stabilization plans for state approval. If they complied with approved plans, mill owners would be released from all liability for the tailings. Enforcement of the stabilization plans was to be the responsibility of the states.[58]

Although the joint statement created an impression of interagency harmony, it was far from being a real compromise on the underlying differences between the AEC and the FWPCA. Almost before the ink was dry on the agreement signatures, agency officials reasserted their original position and distanced themselves from the joint statement. Philip Lee, assistant secretary of the Department of Health, Education and Welfare, summed up his agency's position:

> [It is HEW's] belief that this [joint agreement] is a satisfactory way in which to initiate action, but that it should not be used as a precedent for the handling of this type of waste material [and] any new installations should be located and the material should be processed and disposed of in such a manner that questions of immediate and long-term exposures of people to hazardous materials are minimized. . . . Presently, it is not clear that, after stabilization and containment of the piles, [radiation from the piles] will not still constitute a hazard.[59]

Allen Jones of the AEC's Division of Raw Materials candidly informed the Colorado State Board of Health on December 12, 1966, four days after the joint statement was signed, that the agreement was not to be regarded as a statement of position or as prior approval of future actions. Glenn Seaborg, chairman of the AEC, declared in early 1967 that the AEC saw no reason to change its basic views on the tailings piles that the agency had expressed during the Muskie hearing.[60]

Both the AEC and the FWPCA operated from a different philosophical perspective regarding tailings, but the practical stumbling block in resolving the problem was the failure of any government agency or private company to accept financial liability for long-term tailings stabilization. During the Muskie hearings, Morris of the AEC made the following point: "From a purely radiological standpoint, we do not think we could justify [the FWPCA recommendations], except we should minimize radiation if it does not cost too much or is not too impractical."[61] Colorado health officials were also concerned about the looming costs of any long-term tailings management program. Dr. Roy Cleere, director of public health for Colorado, concluded that long-term stabilization of the piles was a legitimate operating expense of the mills and that the AEC should contribute to remedial action costs.⁶² In the FWPCA's written response to questions submitted to the agency during the Muskie hearing, it stated:

> There have been no requirements for control [of tailings at closed mills] in AEC licenses and contracts. One can conclude that no privately-owned tailings piles have been subjected to control measures because of the fact that control of tailings piles does not create profit. As a result these private companies would have to pay such costs out of their planned profits from uranium extraction. Industry has been reluctant to do so as this would reduce the profit to company stockholders. The costs of interim measures are exceedingly small when compared to the companies [sic] past receipts.⁶³

Throughout the controversy between the AEC and the FWPCA, some uranium mill company officials themselves expressed interest in long-term tailings management. In May 1966, when an AEC official surveyed the twelve milling companies that controlled twenty-five of the thirty-four mill tailings dumps, the millers said that they were willing to take appropriate steps for long-term stabilization of their tailings once it had been made clear to them what steps were required. Page Edwards, president of the Vanadium Corporation of America, commented privately that his company would take action on its tailings if it could be assured of firm criteria and guidance by health officials. Representatives of the Kerr-McGee Corporation concurred with the established AEC position that their tailings did not pose a radiological threat but pledged that it would cooperate with health agencies and the AEC and do anything necessary to protect the public health. The Atlas Minerals Company "displayed a great deal of initiative in handling the tailings problem." Atlas planned to allocate money each month, beginning in 1966, to build a reserve of capital to stabilize its tailings piles gradually at its Moab mill. Atlas president Roy Hollis believed that mill owners, if given reasonable guidance, would choose to take care of the problem of long-term tailings management by themselves. Most industry officials, however, took little practical action on their own to begin long-term tailings management on the level contemplated by public health officials. Millers were reluctant to undertake any management program without clear leadership from either state or federal agencies.⁶⁴

All seven Colorado River Basin states concurred unanimously with the conclusions and recommendations of the FWPCA report and continued to hope that they would be able to rely on federal agencies to assume leadership for tailings management programs. Although the joint statement signed by the AEC, the FWPCA, and the USPHS purported to create a framework for solving the tailings problem, the obvious lack of federal agency commitment to the framework reinforced state and industry fears that they could not rely on federal agencies to undertake long-term tailings management programs or offer financial support for stabilization measures. Because the AEC placed the burden of tailings management on the uranium industry's shoulders, the industry asked the Colorado Board of Health to consider adopting regulations governing inactive uranium tailings piles. Responding to that request, the department announced that it would take the lead in proposing and planning tailings management strategies for mill owners operating in Colorado. The uranium millers agreed to comply with the planned state regulations.[65]

The Colorado Department of Public Health convened a hearing in May 1966 to solicit public comment on its proposed rules and regulations to control the disposition and storage of abandoned uranium tailings. Colorado officials were "pretty much feeling their way" at the time, and they hoped to consult with the USPHS to establish criteria for evaluating the mill stabilization proposals. Responding to the urgent concern of Colorado health officials, the State Board of Health, within days of the public comment hearing, adopted a preliminary version of a new regulation requiring mill owners to submit to the state their plans for tailings stabilization. The regulation required mill owners to cover their piles and stabilize them against erosion, control water runoff from adjacent lands flowing onto the piles, limit access to the stabilized piles, and maintain their piles "in such a manner that excessive erosion of . . . or environmental hazard from radioactive materials does not occur."[66] The regulation also required that "[t]he stabilized pile . . . be surrounded by a chain link fence six feet high which shall be posted conspicuously with a sign or signs bearing the words 'Keep Out, Radiation Area.'"[67] The regulation was finalized in December 1966. Colorado was the first state to exercise its authority. Several states followed Colorado's lead, with the Colorado regulation serving as a model for subsequent state legislation.[68] The AEC had finally succeeded in forcing primary responsibility for abandoned tailings onto industry and the states. On a practical level, it was a victory for the AEC. More significantly, however, the Colorado regulation reflected the concerns raised in the FWPCA report and validated the environmentally conservative position that would dominate subsequent policy battles over tailings.

Although public health authorities in Colorado and other states took an active role in tailings management, many federal policy makers still believed that it was ultimately the national government's responsibility to solve the tailings pollution problem. Congressman Wayne Aspinall (D-Colorado), longtime member of the Joint Committee on Atomic Energy and chairman of its Subcommittee on Raw Materials, took a particular interest in the mill tailings issue because several tailings piles were in his district. In addition, he was a powerful political ally of the uranium industry and a keen observer of governmental and environmental matters. In early 1967, in letters to AEC chairman Glenn Seaborg and John Gardner, secretary of Health, Education and Welfare, Aspinall expressed his fears about the tailings:

> I am concerned . . . that after several years of probing for answers, there is still less than complete agreement on the seriousness of this problem, or on the measures which should be employed to cope with it. . . . Regardless of specific regulatory responsibilities, the national need which has generated the mill tailings situation implies a federal government obligation to take the lead in working toward a resolution of any and all problems which result. . . . My personal feeling is that any further confusion or delay in resolving this matter is an injustice to those people who may be personally affected, and places an undesirable stigma on our atomic energy programs in general.[69]

Aspinall was not alone in his growing uneasiness about the tailings. Ultimately, it would be just that sense of equitable justice that compelled Congress to address the tailings pollution problem a decade later.

Perhaps the AEC's greatest shortcoming during the uranium mill tailings water pollution controversy was its failure to appreciate the readiness of ordinary citizens to accept the views of health experts on the issue. As the tailings problem grew from a regional issue to become part of the nationwide environmental debate during the 1960s, the AEC's resistance to well-founded criticism of its handling of the tailings issue increased proportionately. The result of the AEC's reluctance to acknowledge the potential long-term environmental risks posed by uranium tailings did not reassure the public about the AEC's commitment to shield the public from atomic energy hazards. On the contrary, it helped to undermine the public's confidence in the Commission and the nation's atomic energy program. The agency's technical and legalistic responses seemed increasingly suspect to the growing numbers of Americans who were becoming sensitive to issues of environmental pollution. The agency, responsible for the promotion and regulation of nuclear power, should have known its behavior was likely to initiate a serious public relations disaster whether or not it had the law on its side. At the very time when the AEC needed to take decisive leadership and action to manage atomic pollution that respectable public health experts feared might be a potential long-term hazard, it looked the other way and pressured states to shoulder the burden of managing the tailings.

At least with respect to water pollution, the AEC's efforts to control effluent from operating mills and the states' activities to limit the erosion of the abandoned piles provided a temporary solution to the crisis and slowed the rate of further environmental degradation. The mountains of tailings remained, however, and the most lasting impression the public took from the mill tailings water pollution controversy was the AEC's reluctance to safeguard public health. That unfavorable opinion would haunt the AEC in all its future dealings with uranium mill tailings.

5

Warm Air
Tailings and Air Pollution

THE LAST FEW MONTHS of 1966 appeared to be a time of respite in the debate over uranium mill tailings policy. Radiation water pollution in the Colorado River Basin had been reduced and a monitoring program kept close watch on basin water quality. The most hazardous tailings piles had been pulled back from the edges of rivers or were in the process of being moved. The public health officials and the uranium millers had finally accepted the AEC's position that tailings management at abandoned mills was the responsibility of the industry and the states. Colorado established tailings management regulations, and uranium millers in the state were developing programs for long-term stabilization of their tailings. The outcome of the water pollution dispute delighted AEC officials, who believed that "the regulations adopted by Colorado which require stabilization of uranium mill tailings appear to be an excellent approach in solving the uranium tailings problem."[1] Following Colorado's example, other state governments developed strategies to cope with tailings pollution, including establishing timetables for enforcement of new state regulations that required millers to establish tailings management programs.[2]

Because the radium water pollution problem appeared to have been resolved, the Federal Water Pollution Control Agency's role in the tailings issue waned. FWPCA experts reported that the tailings were not an immediate threat to the water supply, and once the tailings were stabilized to prevent erosion there would be "no concern" about long-term radium water pollution.[3] The interagency conflict between the AEC and the FWPCA was settled, at least superficially, by the agencies' joint statement issued in December 1966. All the agencies hoped their agreement would serve as the basis for cooperation and good relations between the AEC and federal and state public health agencies. The AEC considered it the conclusive statement of tailings policy. Nevertheless, the joint statement did not completely erase the simmering uneasiness among public health experts about tailings safety. In addition, some legislators and state health officials remained suspicious about the AEC's commitment to atomic energy safety and environmental quality because of its steadfast refusal to assume a leadership role in resolving the tailings water pollution problem. According to these critics, the AEC failed to meet its obligation to safeguard the public from atomic hazards. Atomic technocrats inside the AEC, however, were confident that the tailings problem had been laid to rest. Once again, they underestimated the scope and volatility of the uranium tailings issue.[4]

Water quality in the western rivers improved once the AEC enforced its radiation protection regulations at the mills. Although public health experts remained vigilant about radium water pollution, they now turned their attention to the problem of atmospheric pollution created by the milling industry. Radioactive material frequently contaminated the air in mills and blew from the tailings piles to blanket nearby cities with a fine layer of potentially hazardous dust. Health officials feared that this dust, and the radioactive radon gas emanating from both stabilized and unstabilized piles, might endanger nearby communities. The stage was set for another confrontation between the AEC and its critics.[5]

Beginning in the late nineteenth century, European scientists published extensively about the atmospheric hazards of radioactive materials, particularly hazards posed to miners. English translations of several of the most influential European studies were widely available to American occupational hygiene experts before World War II. By the mid-1940s scientists clearly understood that radioactivity of uranium ore and the radon gas in the mine air caused lung cancer in miners. A few years later, they understood that the radon daughters were the primary culprits. When the great uranium boom erupted in America in the late 1940s, health professionals had little doubt that atmospheric radiation, especially radon gas, could be deadly.[6]

Today, the process by which radon causes lung disease is well known. Radon gas is an odorless, tasteless, colorless gas produced by the uranium decay cycle. It has a half-life of 3.8 days. When uranium ore is deep underground, most radon gas decays into nongaseous radionuclides before it can migrate to the surface through the rock and soil. Radon that reaches the atmosphere is diluted by ambient air to extremely low levels in most cases. Mining and milling, however, brings uranium ore, and its constituent radium, to the surface and produces a concentrated source of decaying radium. The sandy tailings enabled radon to percolate through the loosely consolidated mill waste more readily than it did through solid ore or densely packed soils. Under those circumstances, radon easily enters the atmosphere. Radon that is inhaled and exhaled while in its gaseous state causes little damage to lung tissue. If the radon decays while inside the body, however, the decay process produces radioactive decay products (daughters) and ionizing radiation. The most damaging form of radiation from radon is the heavy alpha particles that can disrupt the structure of lung cells, increasing the risk of cancer. The radioactive radon daughters attach themselves to small particles, such and dust and smoke, which can remain in the lungs and continue to emit radiation as they decay. Atmospheric radon can travel hundreds of miles before it eventually decays to solid, nonradioactive lead 206. Usually, as the surrounding outside air dilutes the gas, radon levels dissipate quickly to nearly background radiation levels within a few hundred yards of its source. Inside closed areas such as mines and mills, however, radon can concentrate to dangerous levels. Radioactive dust can blow considerable distances and continue to emit radiation for as long as the radionuclide in the dust decays.[7]

Uranium mills were as dirty and dangerous as any large-scale ore reduction operations. They exposed workers to a wide range of industrial hazards, including heavy machinery, caustics, and acids. Mill employees also handled mountains of rocky ore. The AEC reported that as of the end of 1957, the ore feed rate at the twelve operating mills was ten thousand tons daily. Unlike other mills, uranium operations posed unique risks. Among the most dangerous locations in uranium facilities were the crushing houses where the ore was pulverized, areas where radioactive materials were transported or stored, and rooms where the powdery yellowcake was packed for transport. It was there that dangerous radioactive materials and radiation were most likely to escape into the atmosphere and contaminate workers. During the early days of uranium milling, workers inhaled rock dust, radioactive particles, and radon in the crushing house. The dust was a source of silica, as well as radioactive contaminants, that increased the risk of silicosis in mill workers. Careless transport and storage practices failed to contain radon and dust. The yellowcake was packaged in drums, and in the packing rooms the flourlike powder permeated the atmosphere. Yellowcake was a chemical respiratory health hazard to mill workers because the dust particles retained some of the acids or alkaloids used in ore processing. Mill workers risked chemical toxicity to their internal organs, especially their kidneys, and nonmalignant respiratory disease. The yellow dust was also a radiological respiratory health hazard because the dust particles contained radioactive uranium decay products. Contamination spilled out of many mills, too. During the heavy milling in the 1950s, uranium operations released into the atmosphere considerable concentrations of radioactive decay products of uranium, including radioactive isotopes of uranium and lead, as well as radium and thorium. The small dust particles easily bypassed natural respiratory filters (such as the nose) and lodged in the lungs, where the radiation emitted from the dust caused tissue damage and increased the likelihood of lung cancer in people who inhaled the mill air pollution.[8]

Indications that the atmosphere in the mills might expose workers to radiological hazards first appeared during the war, but most atomic scientists, mill owners, and mill workers were not concerned. Ralph Batie was concerned. As a Manhattan Project scientist (later head of Health and Safety for the AEC's Colorado Raw Materials Division) in charge of occupational safety at the wartime uranium mills, he was aware of the studies on European miners and reasoned that similar conditions might exist in the uranium mills of the Colorado Plateau. Batie's brief mill investigations showed they were extremely dusty. "When we first started in the mills," Batie testified years later, "the Army got a little concerned. . . . We were mainly interested in dust. . . . [I]t was flying everywhere."[9] He feared that conditions in the mills might lead to a cancer epidemic among mill employees.

Batie found it difficult to convince mills to implement dust abatement programs. Companies were reluctant to spend money for safety measures that the army did not require. The mill workers themselves balked at Batie's solutions, such as dust filter masks, as inconvenient and uncomfortable. Based on the evidence on radon in mines, it appeared likely that conditions in the mill were less acute than in the mines and that mill workers were not exposed to as high concentrations of radon in small spaces as were uranium miners. Finally, although scientists knew that radiation damaged living tissue, they were uncertain about radiation tolerance levels. European epidemiological studies proved statistically that atmospheric radiation in confined spaces increased the likelihood of cancer, but they did not offer explicit guidance to experts interested in safeguarding the health of uranium mill employees. American investigators, like Batie, also had problems accurately assessing the condition of the mill atmosphere with the primitive measuring equipment available to them at the time. Under the circumstances, managing possible health hazards at uranium mills remained a low wartime priority.[10]

After the war, when the converted vanadium-uranium reduction mills closed, questions about mill atmosphere safety were set aside. However, when uranium companies reopened the mills and built new ones in the late 1940s to meet the surging postwar uranium demand, Batie and other public health experts renewed their concern about contamination caused by blowing and accumulating dust in and around the mills. Armed with new, more sensitive measuring equipment, Batie uncovered conditions that made him even more apprehensive than he had been a few years earlier. The floor of one older mill continued to emit radiation three years after its postwar closure. "At some of the more dusty operations in the mills...," Batie testified, "the readings indicated that severe internal radiation hazards existed in many operations." The Climax Uranium Corporation mill in Grand Junction, originally a sugar mill, was still being used, in part, to store sugar years after most of the structure was converted for uranium milling. A Colorado state health inspector noted with alarm that the uranium dust on the sugar sacks was so thick that he could write his name on them with his finger.[11]

Batie and regional AEC Health and Safety Division officials warned their superiors and the Colorado Department of Public Health that radiation contamination at the uranium mills might be a health problem. While they strongly suspected that the mill conditions were dangerous, they were hindered because of the lack of evidence to support their conclusions. In 1948 the Commission assigned Dr. Bernie Wolf, a respected radiologist and director of the AEC's medical division, to investigate mill conditions. He reported his findings in a memorandum to L. C. Leahy, manager of the AEC's Colorado Area Office. He was shocked by the "slipshod practices" in the four operating mills he examined and outlined his concerns about the dangerous conditions. First, he observed that some workers experienced short-term upper respiratory tract irritation, called "vanadium hack," that could lead to chronic lung damage. Next, Wolf noted that silica dust levels were dangerously high. At three of the mills, the dust was between 41 and 49 percent free silica and sat one to three inches thick on the floors and rafters. Most troubling for Wolf was his discovery that radioactive dust levels were "excessive." He recommended that comprehensive health monitoring of workers be undertaken, that "attention should be directed towards the matter of control over waste materials and the spread of radioactive dusts to areas adjacent to the mills," and that mills should provide better ventilation and changing rooms to limit the spread of radioactive dust off the mill site.[12]

Wolf's suggestions attracted little attention in Washington. In contrast to the conservative approach of state and other federal health officials, the AEC was convinced by the available data that the tailings did not pose an "unreasonable risk to public health and safety," and it was not greatly concerned about possible long-term harm from tailings dust.[13] In the agency's view, it was unlikely that individuals in the immediate vicinity of a tailings pile would be exposed to dangerous concentrations of airborne radioactivity. "The physical and chemical characteristics of uranium mill tailings," the AEC asserted, "do not differ materially from those of non-uranium mill tailings, except that the concentrations of naturally occurring radioisotopes in uranium tailings are somewhat higher."[14] The AEC thus sought to minimize any health concerns by asserting that the only difference between uranium and other processing waste was that uranium tailings were more radioactive. Ironically, this was the precise information that caused state and federal public health officials to fear that the tailings *might* pose a potential long-term health risk.

Taking matters into its own hands, on May 9, 1966, the Colorado State Board of Health ordered uranium millers to submit plans for stabilizing their tailings piles by the following autumn. The state, along with the health departments of Utah and Wyoming, also asked the Commission to effect stabilization requirements and not to terminate the licenses of mill companies until they brought their tailings under control. The states suggested that stabilization costs were a mill operating expense and that the AEC, by contracting for uranium, was also responsible for them. The AEC responded by downplaying the potential for harm. In addition, it claimed that the results of the 1960 National Lead Company test, which strongly hinted at a potential exposure problem, could not be used as valid evidence to show that tailings were hazardous. The Commission insisted that the test had been designed merely to develop analytic techniques. Other investigations, it asserted, did not appear to show a sufficient basis to warrant AEC license control. Finally, the Commission maintained that although it exercised licensing authority over operating mills for environmental purposes, the fact that tailings contained less than the 0.05 percent uranium or thorium content and were therefore no longer "source materials" meant that the Commission was legally prohibited from requiring millers to bury or dispose of the tailings at closed mills to prevent wind erosion.[15]

Outspoken local critics were not persuaded by the Commission's legal arguments and questioned reassurances from the agency and the milling industry that the yellow-gray film of tailings dust was not a serious health hazard. They fueled a constant sense of doubt among the mill town residents.

One antagonist, Chester Wigton, a physician in Durango, told the Colorado State Board of Health, "The present population will suffer directly from increased risk to radio active [sic] air pollutants. . . . You can't prove it, but I feel anytime you have dust in the air you increase the risk of bronchitis, bronchiectasis, emphysema, and cancer of the lung."[16] Wigton, Dr. Dean Furry, president of the La Plata County Medical Society, Charles Gordon, president of the Durango Chamber of Commerce, and Dr. B. J. Bardin, a local radiologist, captured the sour public mood in their letter to the Colorado Department of Public Health, the AEC, and Congressman Wayne Aspinall: "It is our opinion that [VCA] will be liable for a definite increase in Bronchiectasis, lung cancer, Leukemia, etc. in the citizens of Durango. Only time will bear this out and by then, as you know, it is way too late. . . . We again plead that *immediate action* be enforced on V.C.A. regarding this matter." Other Durango residents were more blunt, accusing VCA of "stalling" and of "wanton negligence to the physical health of the citizens of Durango."[17]

Faced with increasing pressure from state health departments to take a decisive role in managing atmospheric pollution from tailings, as well as with outspoken criticism from Wigton and other community leaders, the Commission considered the need to develop a plan to deflect the growing public disapproval of its hands-off approach to the dust problem. Some AEC officials even believed that the agency had an obligation to mitigate damage to communities caused by uranium tailings since the tailings were a direct result of AEC procurement activities. "In the interest of good government-public relations," wrote Rafford Faulkner, director of the AEC's Division of Raw Materials, "the AEC should use its best efforts to help solve the [dust] problem in cooperation with the local communities and encourage the companies to make reasonable efforts to eliminate the problem." Consequently, the agency requested that milling companies not abandon their tailings piles near communities and intensified its program of encouraging mill owners to manage their tailings piles more effectively. To that end, it provided technical data and advice to the industry and the states on methods of stabilizing the abandoned tailings piles. Nevertheless, except in the mill towns, the more pressing problem of interstate radium water pollution overshadowed atmospheric tailings pollution in the Colorado River Basin.[18]

While the AEC and state officials worked to persuade millers to limit wind erosion of their tailings piles, some public health experts shifted the focus of their attention from the blowing dust to the more ethereal, and possibly more dangerous, problem of radon gas emitted directly from the tailings. In 1966 the Colorado Department of Public Health and the USPHS became concerned that even if the tailings piles were stabilized to stop wind erosion, the piles would continue to emit radon in sufficient quantities to threaten public health. Although health experts knew that radon in high concentrations in closed areas, such as mines, could be fatal to people who breathed it, they were less certain about the potential health hazards from radon exhaled by tailings piles into the atmosphere. The fundamental question was the extent to which the radiation emitted was a public health problem, given the no-threshold radiation exposure theory used by the NCRP, the ICRP, and federal and state agencies. "[I]t is not clear," wrote Philip Lee, assistant secretary for Health and Scientific Affairs of the Department of Health, Education and Welfare, "that, after stabilization and containment of the piles, the emanation of radon and radon daughter products will not constitute a hazard. Consequently, the Public Health Service is planning to make a further assessment of this aspect of the problem in a number of locations."[19] Similarly, the chief of the USPHS Division of Air Pollution, G. MacKenzie, testified before Congress in June 1966 that radon from tailings piles was a possible health threat and should be studied.[20]

In 1966 the Colorado Department of Public Health and the USPHS conducted tests of air quality near several mill tailings piles to establish the potential for radon gas exposure. Air samples taken near the piles in the past demonstrated that the radon emitted from tailings might increase the environmental levels of radon significantly, but the findings were inconclusive. The USPHS found that samples taken in Grand Junction over the tailings and downwind from the pile exceeded the Federal Radiation Council's Radiation Concentration Guide for atmospheric radiation (RCGa) of 3 picocuries per liter, as permitted under Part 20. The AEC, in its own investigation, found that air samples taken within one-half mile of the tailings showed readings one-third to one times greater than the RCGa at the former AEC-owned and stabilized pile at Monticello, Utah. The AEC advised further sampling using techniques that would provide greater measurement accuracy. Subsequent testing with radon "film badges," a new testing technology designed to gather radiation readings over a longer period, provided only indeterminate data. Nevertheless, the evidence was sufficient for the AEC to determine that radiation values downwind of the piles were insignificant. The testing activity by state and federal agencies also led some in the uranium milling industry, most notably Kerr-McGee, to review the results of their own atmospheric testing programs. Industry executives were relieved to find that their surveys showed that airborne radon levels did not exceed regulatory levels, and, therefore, they, too, asserted that the tailings were not a health hazard.[21]

Nobody was surprised by the inconclusiveness of the test results. Radon levels fluctuate widely, the air samples were taken over a short time, and atmospheric conditions probably interfered with the accuracy of the tests. Researchers noted that meteorological conditions could cause a tenfold difference in air contaminant levels from day to day. The "grab sample" method of evaluating air quality was not accurate because it measured contamination of a small amount of air, from one to sixteen liters, over a period of only a few minutes. The film badges, originally designed as personal dosimeters in uranium mines, proved to be incapable of measuring radon levels less than the relatively high dose of 10 picocuries per liter. All parties agreed that under such conditions assessing radon hazards accurately was difficult and that a more sophisticated, comprehensive, and decisive study was required. Still smarting from the public relations debacle over radium water pollution, the federal agencies agreed to coordinate their investigation efforts with state authorities. The AEC suggested that a collaborative approach between the Commission and the USPHS on future radon testing was necessary to prevent "misapprehensions with the public, industry, state agencies and Congress" that the federal agencies were competing with each other. The Commission also recommended that the agencies avoid releasing conflicting data about any radon problems—a bitter lesson learned by the AEC in its handling of the tailings water pollution problem. The purpose of the proposed cooperative effort was to provide technical advice and assistance to the states and uranium milling industry in evaluating the public health aspects of atmospheric radon gas and tailings containment programs. Yet, despite the Commission's willingness to participate in testing, it never abandoned its position that "from a radiological safety standpoint, licensing control of tailings is not required."[22]

While the AEC and the USPHS negotiated the ground rules for their joint radon testing program, public and congressional pressure increased to force the Commission to take affirmative action to solve the atmospheric pollution problems. Congressman Aspinall, chair of the Subcommittee on Raw Materials of the Joint Committee on Atomic Energy, was especially concerned about the safety of the piles because many of them were in his district. He gently prodded the AEC and the USPHS to undertake the joint comprehensive study of the radon gas problem.[23] In a letter of February 1967 to Glenn Seaborg, Aspinall outlined his concerns:

> I understand there was still a considerable degree of uncertainty regarding the extent of the hazard (be it short or long range), difficulty in interpreting recent data from AEC and PHS air samples taken at the sites, and continued uncertainty regarding proper methods for stabilization of the piles.... I am making a personal request to the AEC and the HEW to work together in preparing a plan of attack on this problem which will eliminate the seeming confusion that prevails at the moment. In particular, I want you to call to my attention any legislative action which you believe to be essential toward arriving at a solution of the problems related to existing mill tailings, as well as those to be generated by our future requirements.[24]

The AEC responded to Aspinall's appeal by pledging to provide "unequivocal guidance" to the states, but "the states should have primary jurisdiction [over tailings at closed mills] with technical support and advice from the Federal agencies."[25]

State governments urged the federal government to resolve the atmospheric radon problem, too. In June 1967, five months after the Colorado mill tailings regulations became effective,[26] the Colorado State Board of Health conducted public hearings on VCA's proposed plans to stabilize its abandoned piles at Durango and Naturita. The following month, at the Sixth Session of the Conference in the Matter of Pollution of the Interstate Waters of the Colorado River and Its Tributaries, the conferees, including representatives from each of the seven Colorado River Basin states, recommended that responsibility for tailings management be established. They also wanted specific timetables for immediate and long-range programs for environmental protection from abandoned tailings, including protection against radon emissions. Although the conference participants focused their attention primarily on water pollution, they also considered the problem of wind erosion of the tailings, recommending that short-term remedial action to prevent wind erosion of the piles at active and inactive mills be completed by the end of 1970. An internal AEC memorandum reported that Commission representatives felt pressured to accept the aggressive recommendations of the conferees though they were contrary to official AEC policies.[27]

Conferees also discussed the larger issue of the cost of long-term stabilization of the tailings piles to control all water and air pollution. The long-standing position of the Colorado Department of Public Health, and one supported by the state representatives and federal health experts at the conference, was that "stabilizing these piles [to prevent erosion] is a legitimate operating expense and that the AEC should participate in paying the costs."²⁸ Conference representatives recommended that containment of the piles was the responsibility of the millers and that "[t]he AEC as a major purchaser [had] to assume its responsibility in this activity."²⁹ When an AEC representative asked for clarification of this language, an FWPCA official bluntly told him that "assume its responsibility" meant that the AEC should pay the tailings containment bill.³⁰

The Commission claimed that it did not have sufficient funds for the large-scale stabilization programs required to contain the pollution from the tailings piles. Since it recognized, however, that it might have some financial responsibility to help the uranium industry resolve the tailings situation, it suggested that some stabilization costs previously paid, or to be paid by millers between 1963 and 1968, could justify the AEC's paying higher prices for uranium in its few 1969 and 1970 contracts. The AEC insisted, however, that it would not underwrite the entire tailings control effort. Neither the states nor the federal public health officials had any plans, or even suggestions, regarding methods for financing long-term environmental protection.³¹

The media followed the conference closely. It had reproached the AEC as early as 1966 for what critics claimed was its failure to solve the problem of radon gas exhaled by the tailings. The allegations stung AEC officials; they believed that their agency had provided considerable technical advice and encouragement to the mill owners and states on stabilization techniques and that the agency had accomplished a great deal in solving all aspects of the tailings pollution problem, including the problems of dust and radon pollution.³² In an official conference statement to the media the Commission asserted:

> [T]he available evidence indicates that reasonable and sufficient efforts have been, and are being, taken to minimize [radiological pollution]. . . . The AEC plans to continue its cooperative effort with Federal, state, and local authorities and with the milling industry to achieve an acceptable and reasonable solution for dealing with the uranium mill tailings.³³

After the conference, the AEC proposed that federal agencies consider the creation of a perpetual care fund financed by an assessment on ore processed by the mills. AEC officials agreed that the agency would consider such assessments as legitimate costs in negotiating future uranium procurement contracts. The management of the fund was to be left to individual states. Officials of the USPHS and the Department of Health, Education and Welfare received the AEC proposal favorably, and it became the basis for additional study on cost allocation of remedial tailings management.[34]

During postconference discussions, state officials recommended that the AEC, the Department of Health, Education and Welfare, and the Department of the Interior develop a joint long-range program for protecting the environment from pollution created by the tailings. Acting on that recommendation, the AEC agreed to participate in the new Interagency Technical Committee on Control of Uranium Mill Tailings Piles. The committee considered a variety of tailings management strategies, including the AEC proposal for a perpetual care fund, removal of radium from the tailings, and disposal of the tailings in pits or old mines.[35] It discovered that some of these management options were impractical and would "require justification not presently available, especially in view of the expected high costs."[36] One expert estimated, for example, that chemical removal of 90 percent of the radium in the tailings would double the cost of uranium.[37]

The seemingly haphazard approach to testing, the variable sampling techniques of previous studies, conflicting test results, lack of consistent information, and contradictory interpretations of data all contributed to the confusion of government policy makers, the public, and industry executives about whether an air pollution hazard from the tailings existed. Responding to the conference recommendations, beginning in 1967 the AEC, the USPHS, and the Colorado and Utah health departments attempted to determine once and for all if a radiological safety problem existed offsite from the tailings piles.

The agencies focused their attention on four piles, the Commission's stabilized tailings pile at Monticello and unstabilized piles at Grand Junction, Durango, and Salt Lake City.[38] Investigators ultimately took a total of 892 air samples, 209 of which were on-pile readings and the remainder from areas in a roughly two-mile radius of the piles. The results showed that annual radon concentrations directly over uncovered piles ranged between 7.2 and 16 picocuries per liter, including background. Radon levels off the pile diminished rapidly, and although a handful of samples in Grand Junction showed radon concentrations up to 4.1 picocuries per liter in areas up to one-half mile from the pile, most radon levels were far less. At the covered Monticello pile, readings were significantly below those of the uncovered piles, suggesting that stabilization made a demonstrable difference in reducing radiation emissions. By comparison, the Federal Radiation Council RCGa placed the upper limit for radon at 3.0 picocuries per liter, averaged over a one-year period, for areas accessible to the general public. However, researchers noted, the physical characteristics of the piles made it unlikely that anyone would continuously occupy the tailings themselves and receive a radon dose exceeding the guidelines. The study also showed that in inhabited areas farther than one-half mile downwind from the piles, the tailings had not caused elevated levels of atmospheric radon gas beyond the regulatory limits. The agencies concluded that no significant radiation exposure to the public had resulted from the elevated radon levels.[39] One Colorado state health official noted:

> The study was very extensive and . . . conclusively answered the question of whether there was a threat of radon gas contamination of the environment attributable to uranium mill tailings. . . . [T]his question has been raised over the years, but no definitive data was available to even estimate the contribution of the piles to the community. . . . We definitely feel that the negative results obtained, in light of current standards, will help obliterate the fears of the people in the communities that were studied.[40]

The AEC staff was so convinced by the data that they were "unaware of any current problem or development relative to uranium mill tailings with which the Commission should be concerned."[41] As noted in the final report of the study issued in March 1969:

> Atmospheric concentrations of radon in areas near uranium tailings piles have been evaluated as an index to radiation exposure of the population. The results indicate no significant radiation exposure to the public from this source. . . . Development of . . . recommendations to control public exposure to radon from uranium tailings piles is not necessary, as no significant public exposure was indicated by the results of the study.[42]

After so many years of uncertainty, state and federal health officials and the public were relieved that the radioactive dust and radon from the mill tailings piles had been determined not to be significant health hazards. The dust, while a nuisance, could be controlled and "substantial progress" had been made by uranium millers toward voluntary control of uranium tailings piles to reduce the levels of blowing dust. Radon gas levels, although high directly above the tailings, quickly dissipated to near background levels within a few hundred yards. The evidence from all the state and federal studies taken together finally showed that health concerns about dust and radon emanating from the piles had been largely unwarranted.[43]

The problem of atmospheric tailings pollution raised a more significant issue, however, than whether the tailings were hazardous. The differences between the responses of health experts and the AEC to the possibility of hazards fueled growing public unease that the atomic agency was at odds with health officials over safety concerns in the front end of the uranium cycle. Throughout the debate, public health researchers displayed their usual caution in evaluating the potential hazard and were assertive in recommending solutions. As with radium water pollution, state and federal public health agencies led the way in studying and resolving the troublesome issues of tailings dust and radon gas contamination. America's principal atomic energy authority and quasi-independent federal agency most responsible for atomic safety did not display an equally forceful and conservative response to a potential radiation health crisis. Although it participated in studies, the AEC repeatedly pointed out that it was constrained by law and regulation from directly controlling the tailings. Yet those legal constraints reflected the agency's underlying attitude toward front-end hazards. When the problem first arose, the Commission did not deny that a hazard might exist but maintained that it was not responsible for dealing with the problem if it did exist. It concluded earlier in the investigative process than did public health experts that the tailings posed no immediate atmospheric threat to mill communities. It made little effort to work creatively within its legal and regulatory restraints to address the concerns of the public. For the AEC, atmospheric pollution from mill tailings piles was a minor problem from the time it was first raised as a possible health concern in 1958 until the issue was finally resolved in 1969. For the public, the AEC's seemingly reluctant response fueled growing public suspicion that the atomic agency was more interested in atomic development than public safety.

In hindsight, the AEC was perhaps justified in its reluctance to undertake serious and expensive abatement measures for what turned out to be a minor pollution problem. Decades of extensive research has proved a strong link between uranium mining and lung cancer, but no new evidence has been presented that seriously contradicts claims that tailings in isolated areas do not pose a significant health risk. Yet some tailings were located in or near urban areas, and during the 1950s and 1960s, health experts were not convinced that such tailings would not be a hazard. Given the preliminary data that pointed to potential danger to workers and the public from tailings dust and radon, the AEC should have assumed that a hazard might exist and exercised leadership to investigate the threat and protect mill workers and the public until the evidence had established that mill operations posed no significant atmospheric pollution health threat. It chose instead not to make airborne mill tailings pollution a priority. As with tailings water contamination, in handling the issue of atmospheric pollution, the AEC gambled with the health of mill workers and the public. This time it won. The next time the AEC rolled the dice, it was not so fortunate.

Sheet music cover for the tune "Radium Dance," written by Jean Schwartz and published in 1904 by Jerome H. Remick in Detroit, Michigan. A song like this exemplifies America's early romance with radium. Courtesy of the Rare Book, Manuscript, and Special Collections Library, Duke University, call number A-5676, music.

Assayers work in a lab at the Radium Company of Colorado mine in Colorado, ca. 1920–1930. Note the lack of protective clothing and other protective measures. Photograph by Rocky Mountain Photo Company, courtesy of the Western History Department, Denver Public Library, call number X-61278.

View of Primos Chemical Company mill in Vanadium, Colorado, ca. 1914. Photo courtesy of the Western History Department, Denver Public Library, call number X-62302.

Former government vanadium mill in Monticello, Utah, acquired by the AEC and refitted to recover both uranium and vanadium. It operated during the years 1948–1962, producing 4,593,028 pounds of uranium oxide. Photo courtesy of the Department of Energy Collection, Loyd Files Research Library, Museum of Western Colorado, Grand Junction, call number 1983.63 #110.

Dupuse Lease, 10 miles north of Edgemont, South Dakota, in the Edgemont Mining District. Pictured here is the Edgemont Mining Company's Virginia C Mine, 1953. Photo courtesy of the Department of Energy Collection, Loyd Files Research Library, Museum of Western Colorado, Grand Junction, call number 1982.171 #202.

View of a vanadium fusion furnace in a refining room, Monticello, Utah, 1950. The worker is wearing ordinary work clothes with no protective equipment. Photo courtesy of the Department of Energy Collection, Loyd Files Research Library, Museum of Western Colorado, Grand Junction, call number 1982.171 #104.

Worker shoveling yellowcake in a refining room at the AEC's Monticello, Utah uranium/vanadium mill, 1950. His face is exposed and there is no wash or filter in evidence. Photo courtesy of the Department of Energy Collection, Loyd Files Research Library, Museum of Western History, Grand Junction, call number 1982.171 #107.

Worker dumping a wheelbarrow of ore from a loading platform into a mine car soon to be moved to the surface at Woody Powell and Walt Moore's uranium mine in San Miguel county, Colorado, 1950. Photo courtesy of the Department of Energy Collection, Loyd Files Research Library, Museum of Western History, Grand Junction, call number 1982.171 #145.

Miss Atomic Energy contest sponsored by the Uranium Ore Producers Association and the Grand Junction, Colorado Chamber of Commerce, May 1950. Left to right: Shirley Riggs; Eugene H. Sanders, president of UOPA; Karen Keeler; Cathryn Gordon; Jo Reva Beane; Barbara Talarico. Photo courtesy of the Department of Energy Collection, Loyd Files Research Library, Museum of Western Colorado, Grand Junction, call number 1983.63 #163.

Ambrosia Lake, New Mexico, 2001. Photo courtesy of the author.

6

Warm Homes
Indoor Tailings Pollution

THE DECADE AND A HALF after World War II was a period of environmental consensus during which issues of natural aesthetics, urban open space, and outdoor recreation dominated environmental policy making. Americans viewed pollution matters, especially those associated with atomic energy, as minor problems that could be resolved by technological ingenuity. One barometer of the national environmental mood appeared in November 1960, when the President's Commission on National Goals issued *Goals for Americans*, which highlighted the issues it believed would be vital to America's well-being during the 1960s. The report presented fifteen goals demanding national attention; resolving pollution problems was not among them. The currents of change that touched society during the 1960s, however, spurred a growing public concern about environmental issues, including pollution generated by the nation's atomic energy program. Within five years after publication of the report, environmental protection leaped to the forefront of public consciousness.[1]

National ambivalence about atomic energy grew despite the best efforts of the AEC, which strenuously reassured the public of the 1950s and 1960s that low doses of radiation emanating from the nation's nuclear infrastructure did not pose significant public health problems. The agency also proudly proclaimed that it placed public safety among its highest priorities. The nation was not convinced. Health experts regularly raised troubling safety questions. Some concerns were legitimate, others were not, but few ordinary citizens had sufficient knowledge about atomic science to distinguish between real and imagined dangers. Despite the passion with which atomic boosters promoted the new technology and the determination with which governmental experts comforted Americans about its risks, they were never able to arrest completely a growing public uneasiness that the new technology was more dangerous than advertised. Ongoing concerns about radioactive uranium tailings helped to fuel the broad suspicion among health officials and the public about the safety of nuclear power. Uranium tailings pollution, which began as a western regional problem addressed primarily by public health officials, scientists, and technocrats during the 1950s and early 1960s, evolved quickly into a matter attracting nationwide public attention and criticism. The flash point of that harsh scrutiny was the contamination of the western-slope community of Grand Junction.

The reaction of state and federal health experts and the public to the absence of a comprehensive AEC policy for tailings compelled the development and implementation of remedial programs to limit tailings water and air pollution. The uranium processing industry, working in cooperation with the states and the federal government, had by the mid-1960s significantly reduced immediate water pollution hazards from tailings piles. Ongoing water monitoring ensured that radium levels remained within acceptable limits. Airborne contamination within the mills had been brought under control. Although subsequent testing refined upward the levels of atmospheric radiation outside the facilities,[2] government officials, health experts, and the public were relieved when test results vindicated the AEC's long-standing assertions that the tailings, while a nuisance, were not a serious atmospheric health threat. By the mid-1960s, AEC officials were confident that the most serious tailings pollution problems had been resolved. Reflecting on the Commission's success in reducing tailings contamination, Peter Morris of the AEC testified on May 6, 1966, before Edmund Muskie (D-Maine) and the Senate Subcommittee on Air and Water Pollution that the agency had

> considered speculation that, because of the long half-lives of the radioactive material contained in the uranium piles, they may become hazardous at some future time. . . . [W]e find it difficult to conceive of any mechanism whereby the radioactive material which is now so widely disbursed could become so concentrated as to exceed current applicable standards for protection.[3]

In fact, less than a year later Colorado health officials discovered that just such a "mechanism" for concentrating radioactive mill tailings had been operating for at least a decade. Since the early 1950s, builders had used radioactive uranium mill tailings for construction in Grand Junction and other western cities and towns. The resulting contamination caused radiation levels to soar in the affected communities, and the problem of radioactive uranium mill tailings pollution again erupted into a national environmental issue.[4]

The uranium processing industry had made progress toward controlling their tailings, but some piles were difficult to manage. The risk of dam failure at tailings impoundment facilities was a constant worry. Frequently the piles were too soft, steep, or unstable to permit heavy equipment to contour and flatten them for stabilization. The industry investigated natural and artificial coverings for the piles, but no ideal solution for safely containing the piles was at hand. As the tailings piles grew larger, some millers feared they would run out of storage space. To help reduce the size of their piles to manageable levels, uranium millers encouraged beneficial uses for their tailings. The sandlike tailings, which were milled to a fairly uniform consistency prior to chemical processing, were an ideal building material. Some millers gave tailings away, relieved to get rid of as much of them as they could. Contractors happily used the free sand in their building projects.[5]

Some public health experts even encouraged construction applications for tailings as a way to reduce tailings hazards. In its 1966 report on tailings water pollution, the FWPCA stated that its recommendations regarding possible long-term health consequences of the tailings were "not intended to discourage or prevent the use of mill tailings." "To the contrary," the FWPCA said, "[we hope] that . . . beneficial uses that incorporate adequate disposition will develop."[6] The FWPCA suggested that tailings would be useful for subgrade material on highways or for landfill. Earl Peterson, of the University of Utah Department of Radiological Health, claimed that tailings wastes at the Vitro Corporation Plant in Salt Lake City were a health threat. Beneficial use, Peterson believed, was one solution. "I have recommended," he wrote, "that these wastes be incorporated in some of the large earth fills for construction of roads. The radioactive wastes if buried as a core in such fills, would be essentially permanently disposed of, and would contribute virtually no radiation to anyone."[7]

These apparently benign uses for tailings actually posed a greater risk than originally believed. The tailings that were removed from the mill piles and used in construction emitted gamma radiation and radon. If the tailings were used for construction projects that were far from human habitation or in which they were buried deep underground, the radiation and radon posed little concern. But tailings used in building homes, schools, and businesses frequently raised gamma radiation and radon levels to the point where they were a potentially serious radiological health concern. The problems arising from mill tailings used for construction are best illustrated by events that occurred in Grand Junction, the city most seriously contaminated by tailings.

The Climax Uranium Company began producing uranium oxide at its mill in Grand Junction in June 1950 and continued operations for almost twenty years. Climax processed between nine thousand and fourteen thousand tons of ore per month, the bulk of which ended up as tailings. Like all uranium mills, the Climax mill was privately owned but was "controlled, sanctioned, and licensed by the AEC," and for more than sixteen years of production, until December 31, 1966, the AEC purchased all the yellowcake the mill produced.[8]

As a "public service," since the early 1950s Climax allowed, even encouraged, builders to haul away its tailings. The practice was established company policy by 1956 and continued for another decade. The motive for Climax's behavior appears to have been genuine goodwill, because the mill received no financial gain from the tailings and had plenty of land to expand its tailings piles even without the removals. The giveaway, moreover, was just what Grand Junction needed for its postwar, uranium-rush building boom. About four thousand structures were built in the greater Grand Junction area between 1952 and 1966, and nearly all of them needed some kind of fill material. In addition to residential and commercial construction, contractors used Climax mill tailings for a variety of other purposes, such as subbase for roads and backfill in sewer ditches. A substantial quantity of tailings was used as ballast in empty ore trucks returning to the uranium mines, where the tailings were dumped on the mine waste dumps. Farmers and gardeners found that the sandy composition of the tailings loosened the tight soil in their fields, aerating the ground and allowing moisture to percolate freely. In addition, the tailings were slightly acidic and helped to neutralize alkaline soils.[9]

No detailed records exist documenting the amount of tailings removed from the Climax tailings pile, but company representatives estimated the figure to be as high as 300,000 tons. Of that total, 150,000 tons went to the Grand Junction Highway Department for subbase and packing around culverts and water and sewer lines. The Denver & Rio Grande Western Railroad used 1,000 tons for construction backfill and ballast for empty cars. The Colorado State Highway Department used 25,000 tons as backfill around culverts for Interstate 70 near Grand Junction and 50,000 tons under Highway 50 south of Grand Junction. Contractors used 3,500 tons as sewer line backfill at a local state training school and 5,000 tons under sidewalks and driveways in the Grand Junction area. Local farmers neutralized their soil with as much as 15,000 tons of tailings. Mining companies used an unknown amount as ballast for ore trucks. About 50,000 tons were unaccounted for, but a large portion of that tonnage was probably distributed under and around buildings throughout Mesa County, in which Grand Junction was the largest city.[10]

State and federal public health officials were in Grand Junction in January 1966 during the investigation of atmospheric tailings pollution, to test a new radon detection apparatus called the "radon film badge." Investigators hoped that the film badge, designed to measure radon levels in mines, could be adapted to measure radon exhaled by the tailings piles. The two scientists, Robert Siek of the Colorado Department of Public Health and Robert Snelling of the USPHS Southwestern Radiological Health Laboratory, noticed local builders using uranium mill tailings to backfill a construction site in the city. Informal inquiries revealed that it was a common practice to use Climax mill tailings as backfill against foundations or under concrete slab floors in new construction. Siek and Snelling placed radon film badge devices inside finished buildings where they knew tailings had been used in construction. The badges were too insensitive to give precise readings, but the results of the test, and subsequent film badge tests, showed that the tailings increased radon gas levels in several Grand Junction buildings.[11]

The potential for radiological harm from tailings used in construction worried state public health officials. They knew that tailings produced both direct gamma radiation and airborne radioactivity from residues of radium and its radioactive daughter products, including radon gas. They also knew that long-term exposure to gamma radiation increased the possibility of developing cancers in people with elevated whole-body radiation doses. Gamma radiation was known, for example, to elevate the risk of leukemia and cancers of the breast, thyroid, pancreas, and skin. Exposure may be reduced only by heavy shielding material, such as lead or layers of concrete that are thicker than most basement foundations. Gamma radiation from tailings under and around homes posed a higher risk to children than to adults, primarily because growing children are generally more sensitive to the negative effects of the radiation than are adults. Fetuses are extremely sensitive to gamma radiation, which can cause genetic damage, especially at critical times in the gestation period.[12]

Health experts were most concerned, however, about radon and its radioactive daughter decay products. Some radon daughters decay quickly, emitting gamma, beta, and alpha radiation, but alpha radiation predominates and poses the greatest threat from tailings. When inhaled, the radon daughters are deposited in the lungs. They are potent alpha particle emitters that can damage lung tissue and increase the risk of chromosomal damage, abnormalities, and lung cancer. Gaseous radon is capable of diffusing through porous basement walls and cracks in foundations and thus accumulates. Once the radon gas is inside the home, it deposits its radon daughters in the living spaces. The concentration of radon in buildings is affected by temperature, atmospheric pressure, porosity of basement walls and floor, and ventilation. During the winter months, most homes are tightly sealed against the cold, limiting ventilation and increasing radon levels. Moreover, people usually spend more time indoors during winter, thereby exposing themselves over longer periods to the higher concentrations of both gamma radiation and radon.[13]

Although state health officials recognized the potential for a serious problem in Grand Junction, they were uncertain about how to best approach the issue of tailings used for construction. In January 1966 alarmed Colorado health officials relayed their concerns about the use of tailings for building purposes to Climax representatives. In response, Climax executives proposed that the state health officials institute control measures allowing builders to use the tailings if they would be isolated from human contact but restricting their use for projects that might pose radiological hazards. Climax requested written permission from the Colorado Department of Public Health to allow contractors to remove tailings for uses that seemed to be safe, such as fill around water and sewer lines and as subbase for road and footing construction. Climax also asked permission to remove tailings to mine waste dumpsites, where they would be covered by mine overburden. Such applications, Climax officials reasoned, would put tailings deep underground and greatly reduce the release of gamma radiation and radon into the atmosphere. Responding to the millers, state health officials at first doubted that their permission was necessary for tailings to be removed for construction, indicated that the proposed uses were an ideal method of disposing of the tailings, and offered assistance to develop a comprehensive tailings management plan.[14]

The state quickly resolved any uncertainty regarding its ability to regulate mill releases of tailings for construction in summer 1966. Ongoing preliminary investigations revealed that tailings located in or near many city buildings and homes raised levels of radiation well above natural background levels. Once the Colorado Department of Public Health realized the potential magnitude of the Grand Junction contamination, on

August 1, 1966, it ordered a halt to further release of tailings to residential and commercial builders without its written permission. On August 25 the Climax mill manager halted the release of tailings for any uses. USPHS health experts, who had been working with Colorado health authorities in Grand Junction, were also worried about the elevated radiation readings in the contaminated buildings. They relayed their concerns to the AEC in mid-1966.[15]

Public health experts thus knew by summer 1966 that several structures in Grand Junction had elevated radiation levels as a consequence of the tailings, although as yet they had only imprecise measurements of the contamination. They feared that many more buildings might be polluted. The first task for them, then, was to determine the extent of the indoor radiation problem. Health experts had for some time been able to measure accurately the relatively high amounts of radon found in uranium mines, but designing instruments sensitive enough to measure the smaller quantity of radon likely to be found in homes built with tailings presented technical difficulties. Researchers at New York University had developed an instrument capable of precisely measuring low-level radon daughter concentrations in air samples taken over a short time, called grab samples, for use in uranium mines. At the request of the Colorado Department of Public Health, they built a duplicate machine and in spring 1967 state health technicians, using the new equipment, confirmed the radon badge results and verified the existence of elevated radon levels in several buildings.[16]

It was obvious to state health officials who reviewed the preliminary test results that the indoor radiation pollution was as widespread a problem as they had feared. They realized that only a comprehensive survey could gauge the full extent of the crisis but did not have the expertise or resources to undertake such a sweeping inventory. Consequently, the state health department asked for assistance from the USPHS. The federal health agency rejected the state's request, in large part because of AEC resistance to the study. In its evaluation letter to the USPHS criticizing the proposal, the AEC implied that the radon in Grand Junction buildings was not extraordinarily high. Commission experts reported that elevated radon levels were to be expected from the naturally occurring radium in the soil and that in-depth investigation of the problem was not warranted. After all, in comparison to other western communities, the natural background radon concentrations in Grand Junction were fairly high, measuring 0.83 picocuries per liter, or about 0.004 WL.[17]

Rather than support the state research program, the AEC sought to put the indoor radon problem in "proper perspective." In an effort to bolster its assertion that the elevated indoor radon levels in Grand Junction were a natural phenomenon, the AEC funded an investigation of radon in areas of Florida and Tennessee that were known to have naturally high levels of radium in the soil. The AEC expected to find in these southern test areas radiation conditions similar to those found in Grand Junction. Instead, it discovered that the highest level of naturally occurring indoor radon in the Florida test sites was only about 1 percent of the highest measurements from Grand Junction sites where tailings had been used as construction materials. When the results of the southern testing reinforced the seriousness of the indoor radon contamination problem in Colorado, and potentially undermined the agency's credibility, the AEC quietly buried the report. The AEC did not publicly release it until 1971, well after it had become obvious to everyone that naturally occurring radium in the soil was not the most significant cause of elevated radiation readings in Grand Junction.[18]

The Commission was reluctant to tackle the tailings issue for several reasons. It suggested that the preliminary findings showed no reason for anyone to be alarmed about the health implications from the radiation levels in Grand Junction. There was no evidence, it said, that exposure to the levels of radiation from tailings used as construction material have produced, or were likely to produce, "observable biological damage to residents." Beyond that, the AEC maintained its traditional position that it had no legal authority to regulate tailings. Its jurisdiction was limited to the protection of public safety from statutorily defined classes of radionuclides, such as source material. Radium, the parent element of radon, was not one of these radionuclides. Moreover, the tailings contained too little uranium or thorium to qualify them as source material. The agency regulated the discharge of radioactive mill effluent into the environment and the accumulation of radioactive tailings at mill sites, regardless of the uranium or thorium content, because it considered those actions to be an integral part of mill operations. The agency claimed, however, that the transfer of tailings, for example to a building contractor, was not an integral part of mill operations and consequently outside AEC jurisdiction. Finally, the agency continued to claim that its only relationship with mills was as the purchaser of uranium concentrate. The contracts did not reserve the right for the government to direct or control the manner in which the mill operator disposed of the tailings. The regulation of tailings, the agency had always argued, was primarily the responsibility of states.[19]

Undeterred by federal resistance to its plan, the Colorado Department of Public Health elected to proceed on its own with a Grand Junction survey. As part of the project, state and local health officials worked with Climax mill managers and local builders to determine the number of structures that might have been built with tailings. Interviews with seventeen area builders suggested that tailings were under at least four hundred structures and that an undetermined amount of tailings sat under streets, driveways, swimming pools, water pipes, and sewers. Radon samples taken at about one hundred buildings in which contractors were known to have used tailings showed that 42 percent of the residences and 62 percent of the businesses showed radon levels significantly above background. At the Pomona Elementary School, radon levels increased overnight to thirty-eight times normal background levels, requiring thorough ventilation of the classrooms each morning and periodically throughout the school day. Despite the growing body of evidence verifying existence of radiation contamination from tailings, the AEC staff reported in December 1968 that they were "unaware of any current problem or development relative to uranium mill tailings with which the Commission should be concerned."[20]

The ongoing state investigation into indoor radiation contamination attracted first local, then national, media attention. When the Colorado Department of Public Health first uncovered the problem in early 1966, it refrained from publicizing its findings for fear of panicking the public before it had time to fully assess the problem. Nevertheless, news of the potential problem reached the press that summer. Many stories were accurate, some were not. The media reported that Colorado public health experts worried that uranium mine atmosphere conditions might be duplicated in a home built with tailings, a well-founded concern that proved accurate in a few Grand Junction structures. The national media also widely quoted Howard L. Kusnetz of the Occupational Health Division of the USPHS as saying that living in a home built with blocks manufactured from tailings would be like living in a "poorly ventilated uranium mine."[21] The public alarm caused by this statement prompted the USPHS to issue a retraction, announcing that the press had misinterpreted Kusnetz's remarks. By then the damage was done, and the retraction went largely unnoticed. The media exaggerated the indoor radon problem, incorrectly claiming that tailings had been used in manufacturing cinder-block bricks and in children's sandboxes. As the magnitude of the problem in Grand Junction became more apparent, press reports grew even more fantastical. Reporters claimed that basement walls fluoresced in the dark, windows sweated because of the radiation inside the homes, and Grand Junction girls glowed in the dark. Despite such imaginative stories, much of the publicity regarding the use of tailings for construction was disturbingly correct and attracted nationwide attention, most of it critical of the AEC for its lack of leadership in the matter.[22]

Grand Junction residents were at first puzzled by the media attention. Uranium mining and processing had turned the quiet western-slope agricultural town into a booming atomic industrial city. Residents had accepted the inconveniences of the nearby uranium mill as the price for progress and prosperity. They were proud of their nuclear heritage and even incorporated a symbol of the atom into their official city seal. After considerable negative publicity, however, Grand Junction residents began to resent the sensationalized reports and "scare" publicity that captured national attention at the expense of the city's reputation. Especially irritating was Grand Junction's label as "America's Most Radioactive City."[23]

The negative press was not confined to the print media. An ABC television newscast on January 20, 1970, gave the impression that Grand Junction was a nearly abandoned ghost town. After the telecast, some concerned relatives called their families in Grand Junction asking when they planned to move. A handful of people did move, but there was no mass exodus. The majority of Grand Junction citizens were, to be sure, apprehensive about the tailings. Above all, they wanted reasonable, simple answers to their practical questions about how the tailings under and around their homes and businesses would affect their lives. To that end, and believing that the federal government would underwrite the cost of any remedial action, most of the residents cooperated patiently with the official survey teams undertaking the intrusive indoor radiation assessment work.[24]

Not surprisingly, it was local realtors who were most concerned about the tailings. Many realtors found it difficult to interest buyers in homes or businesses built on top of tailings. Classified real estate advertisements in local papers often began with the heading, "No Tailings." One bank refused to offer home mortgages until radiation readings had been made. A Veterans Administration loan officer intimated that the VA might not approve loans to buy homes unless the tailings under them were removed, and the FHA considered similar action. Property owners feared that the cost of removing tailings from under their buildings would lower the real value of their properties just as surely as would leaving the tailings in place. Perhaps the most common concern among Grand Junction residents was that there did not appear to be any foreseeable solution to the indoor radiation contamination crisis. Most residents placed the blame for the tailings problem squarely on the AEC, which steadfastly refused to accept responsibility.[25]

Medical evidence available to scientists and policy makers by the mid-1960s indicated that long-term exposure to gamma radiation and radon emissions could cause serious health hazards. Despite disagreement among health experts about how much exposure was harmful and whether mill tailings located under buildings emitted hazardous levels of radiation, they nevertheless approached the problem cautiously. Grand Junction pediatrician Robert Ross, Jr., for example, became troubled in 1969 by what he described as a "vague sense" that something was not right with his young patients. He believed that there was an unusually high rate of birth defects and cancers in Mesa County children and that radiation from mill tailings might be the cause. He brought his concern to the attention of C. Henry Kempe, chairman of pediatric medicine at the University of Colorado, and the two agreed that Ross's observations warranted further investigation. After the doctors alerted the Colorado Department of Public Health to their suspicions, the state requested funding from the AEC to investigate the extent of chromosomal changes and incidents of congenital deformity in Mesa County infants.

Despite endorsement by Colorado governor John Love, AEC chairman Glenn Seaborg personally notified Love that the AEC would not sanction the proposed study because it was not supportable on scientific grounds. The AEC staff alleged that the sample of seven hundred births per year was too small and that the radiation levels emitted by the tailings was so low that a valid statistical study could not be made. Consequently, the AEC scientists claimed the proposed study could not link genetic damage to the uranium tailings. The study, they maintained, would be irrelevant because it could not provide future guidance about how to manage radiation from the tailings. It is true that the small sampling proposed for the Kempe-Ross study had inherent statistical weaknesses and there were legitimate questions raised about its methodology, but the AEC's objection to the investigation fueled public concern that it was unresponsive to, perhaps even trying to cover up, the indoor radiation hazards.[26]

Governor Love disagreed with the AEC's evaluation of the proposed Kempe-Ross study. Believing that the investigation was not only relevant but also necessary, he financed it from a special state research fund. The doctors ultimately made some alarming conclusions. The death rate from congenital abnormalities was more than 50 percent higher in Mesa County than in other Colorado counties. The incidence of such deaths, moreover, was increasing in Mesa County but decreasing in the state as a whole. Cleft lip and palate defects were twice as common, and the birthrate was lower in Mesa County than elsewhere in Colorado. The overall death rate from cancer was noticeably higher in Mesa County than in the rest of the state, and whereas cancer mortality rates were constant for most of the state, they were increasing in Mesa County. Kempe and Ross strongly implied in their report that the cause of these problems was increased amounts of radiation of the type emanating from mill tailings, and they recommended that all tailings be removed from under Grand Junction homes.[27]

The Kempe-Ross findings did not go unchallenged. The Surgeon General of the United States, the AEC, and the newly formed Environmental Protection Agency (EPA) all criticized the study and underscored its inherent weaknesses. The federal agencies characterized the results as simplistic and attacked the study for failing to establish that radiation, from any source, caused the abnormalities and the death rates. They assailed the doctors for implying that the mill tailings caused the statistical anomalies and for recommending that the tailings be removed from under buildings. The AEC Division of Biology and Medicine concluded that even if there had been an increased incidence of congenital malformation, it would have been the result of epidemiological factors other than radiation exposure. The importance of the Kempe-Ross results, however, was not whether they were accurate but that they attracted considerable negative public attention and reinforced the public's belief that the mill tailings under the buildings of Grand Junction were dangerous. The study also highlighted growing public fear that the AEC was not taking the tailings problem seriously. For many, the study showed that only decisive action could solve the indoor radiation problem.[28]

By the late 1960s indoor radiation contamination in Grand Junction had become a priority of the Colorado Department of Public Health and a growing media issue. Other mill towns, such as Uravan, Colorado, and Salt Lake City, Utah, discovered that they too were contaminated by tailings, although Grand Junction had by far the greatest problem. It was obvious even to casual observers that indoor tailings pollution was a widespread and growing problem. For the AEC, the tailings problem was becoming a serious embarrassment that left it vulnerable to attack from its critics. Many environmentally conscious Americans questioned the Commission's legal position and believed that it bore primary responsibility for the contamination of Grand Junction. Public and political pressure built for greater federal involvement in evaluating and solving the problem. The AEC decided it had to be more responsive to the situation. In March 1969, at a meeting between representatives of the USPHS and the AEC to discuss the evidence of an indoor radiation problem, AEC officials agreed to participate in discussions with the Colorado Health Department and the USPHS about the matter. The Commission also offered to loan the state its radon testing equipment left over from the earlier atmospheric dust and radon tests.

Yet, rather than address the problem aggressively, AEC officials at the meeting were especially interested in generating more data about natural background radiation levels in Grand Junction to be used as a baseline for gauging the indoor contamination. Developing such a measurement baseline would be desirable from a technical standpoint, but Commission officials also still hoped to show that natural background radiation levels in Grand Junction were abnormally high, supporting their claim that most, if not all, of the elevated readings in Grand Junction were attributable to natural causes. The Colorado Department of Public Health, although grateful for any Commission assistance, disagreed fundamentally with the agency's approach. Background radiation levels, state officials believed, were not the important issue. "When one considers the occupancy characteristics of a home, compared with those of a mine, and the fact that children are much more susceptible to this type of exposure than adults," wrote Roy Cleere, the department's executive director, "it is difficult to consider delaying definitive studies as well as studies relating specifically to control measures that can be taken to eliminate the emanation of the radon to the interiors of the buildings."[29]

In August 1969 the Colorado Department of Public Health stepped up its study effort in Grand Junction. Researchers continued to pour over mill and construction records to identify contaminated buildings. In some cases, gamma radiation surveys in the immediate vicinity of those structures revealed additional contaminated areas. Some home owners also were asking for detailed screening. To better organize its survey, the department selected 0.01 WL, an amount slightly more than twice the natural background level, as the screening threshold above which more in-depth sampling and study was warranted. By early 1970 the state found that 46 out of 77 of the businesses investigated and 82 out of the 156 residences were above the 0.01 WL threshold. A few cases showed readings over 0.5 WL. The maximum reading in the homes was just under 2.0 WL. In comparison, the standard then in force for regulating uranium mines was 1.0 WL, and the Colorado "Stop Work" level for uranium mines was 2.0 WL.[30]

Concerned by the AEC's hesitancy in the face of credible and growing evidence of a contamination problem in the major city of his home district, congressman and JCAE member Wayne Aspinall wrote Robert Hollingsworth, the AEC general manager, to urge the AEC to take a more assertive role in assisting the Colorado Department of Public Health in further radiation testing in Grand Junction. Aspinall suggested that such assistance might "constitute the proverbial ounce of prevention." Prodding by Aspinall and other members of Congress, as well as the tidal wave of negative public opinion, was too much to resist. The Commission finally determined in February 1970, nearly four years after the indoor radiation pollution problem had first come to light, that its assistance in evaluating the matter was both "desirable and feasible." The Commission also expressed its willingness to assist, but not to lead, efforts to develop remedial action measures. Ironically, America's foremost atomic energy agency was the last of the relevant public authorities to actively participate in a comprehensive indoor radiation evaluation.[31]

By early 1970 the Grand Junction survey indicated that the elevated radiation problem in the city's buildings was widespread. There was no longer any doubt that tailings contaminated a significant portion of the city. The situation became more grave with every passing day. By December 1970, 1,357 buildings were confirmed as having tailings under or adjacent to them, and an additional 626 locations had tailings on the property but away from occupied areas. Less than one year later, 2,421 structures were confirmed as having tailings under or against foundations and 1,457 properties as having tailings on the property away from occupied areas.[32]

To establish a coordinated and comprehensive strategy to solve the problem, the Colorado Department of Public Health, in January 1970, asked the AEC and the USPHS to develop exposure and remedial action guidelines for indoor radiation exposure. They hoped that clear standards would improve their ability to assess the risks. In particular, state health officials wanted the guidelines to take into account the likelihood that children would be exposed to radon in contaminated homes.[33] Six months later, Surgeon General Jesse L. Steinfeld issued radiation exposure guidelines specifically applicable to homes built with or on uranium tailings. Relying on Federal Radiation Council and International Commission on Radiation Protection standards as a guide, Steinfeld recommended three graded actions for contaminated structures, depending on the radiation exposure levels.[34] If contamination readings were above 0.1 milliroentgen per hour for gamma radiation and 0.05 WL for radon daughters, remedial action was recommended. If the readings were below 0.05 milliroentgen per hour for gamma radiation and 0.01 WL for radon daughters, no action was required. If contamination levels were between the upper and lower limits, remedial action "may be suggested." These levels for exposures were in addition to Grand Junction's natural background radiation of about

0.014 milliroentgen per hour for gamma radiation and 0.004 WL for radon daughters. The higher gamma radiation levels of 0.1 milliroentgen per hour and 0.05 WL for radon daughters would result in maximum annual exposures of 900 milliroentgen and 2.5 WL under conditions of continuous occupancy of contaminated homes. The EPA estimated that such exposures might increase normal expectancy of leukemia by 36 percent, other neoplasms by 9 percent, and lung cancer by 5 percent in Grand Junction residents living in contaminated buildings. The Surgeon General confirmed that the EPA's figures were the minimum estimates for these types of cancer and suggested that such exposures might even double the risk of leukemia and lung cancer. "All values above [the higher exposure levels] would indicate the necessity for remedial action," reported the Surgeon General, "since at these levels the maximum annual exposures recommended by the Federal Radiation Council and the International Council on Radiation Protection for an individual member of the public is exceeded."[35]

Although the Surgeon General's guidelines were a positive step toward developing a plan to solve the indoor radiation problem, they were too vague to provide practical guidance for policy makers. Officials within the AEC were particularly concerned that the recommendations did not define what "remedial action" needed to be taken or, more significantly, who should pay for it. Perhaps anxious about a rush to judgment on the issue of liability for cleanup, Commission administrators did not enthusiastically endorse the Surgeon General's criteria for remedial action. They suggested instead that the indoor radiation investigation continue and that remedial action strategies be discussed at greater length by the federal and state agencies after they knew the full extent of the problem.[36]

With several state and federal agencies participating in the Grand Junction survey, and some pulling in different directions, it became imperative that they coordinate their activities to achieve a timely resolution of the crisis. Consequently, in August 1970 the agencies established the Interagency Steering Committee to advise and assist the Colorado Department of Public Health in identifying contaminated buildings and determining the necessity and cost of remedial actions. In general, the steering committee oversaw the efforts of the state and federal agencies. It included two representatives each from Colorado and the AEC. Because most of the duties of the USPHS regarding indoor radiation had been shifted to the recently formed EPA by that time, the EPA also had two representatives on the steering committee, whereas the USPHS had only one representative. The steering committee ultimately met seven times to coordinate the state and federal agencies' actions and allocate additional funds for the Grand Junction survey.[37]

The superficial cooperation among state and federal agencies to study the indoor radiation problem did not dispel public suspicion about what it saw as the AEC's belated participation in the Grand Junction survey process and its tepid response to the crisis. The Commission continued to be viewed as an agency unconcerned about matters of tailings radiation safety. The barrage of negative publicity was becoming a public relations disaster for the AEC. According to one Colorado newspaper, the agency's behavior threatened to become a "bloody, running sore spot in AEC credibility."[38]

The flood of such hostile criticism and the disturbing evidence of the widespread tailings contamination in Grand Junction was, by the early 1970s, so overwhelming that even officials in the federal atomic energy bureaucracy questioned the advisability of the AEC's official position. Rafford Faulkner, director of the Commission's Division of Raw Materials, was among them. He wrote that although the AEC was not liable for the indoor radiation problem, the federal government should accept some responsibility for alleviating the problem because a significant amount of tailings had been generated to meet the government's demand for uranium. Gerald Fain, a professional staff member of the JCAE, recommended that a change in the AEC's response to the indoor radiation problem "might be desirable." Some AEC officials suggested that the Commission could stop the erosion of its reputation by offering technical advice and financial assistance to owners of contaminated property. "The Commission will bear the brunt of the criticism for the tailings situation," the AEC staff reported, "and should take steps to show the people of Colorado that it wants to help them." Other staffers strongly implied that the AEC had a moral responsibility to assist in alleviating a problem.

Some of these federal dissenters also acknowledged the validity of the argument, repeatedly made by state governments and municipal leaders, that the federal government should accept responsibility for the indoor radiation problem because the tailings had been generated as a result of government procurement policies that benefited the whole nation. It seemed unfair that the mill town home owners should bear a disproportionate cost of the country's quest for uranium independence. These federal officials also recognized the increasing difficulty, if not impossibility, of trying to persuade the public that the indoor radiation problem was entirely the responsibility of the millers, the home owners, and the states. Some AEC staff members noted that although the removal of all the tailings was undesirable because of the high costs involved, it was equally unacceptable to leave the tailings in place. They proposed that remedial action should be undertaken if the health benefits could be maximized while keeping the overall negative economic and social impacts on the community to a minimum. The AEC even sent the Office of Management and Budget (OMB) a proposal outlining the nature of the indoor radiation problem and asking it to comment on proposed coordinated federal action supporting remedial measures in Grand Junction.[39] In a July 1971 letter to George P. Schultz, director of the OMB, AEC chairman Glenn Seaborg sensed the public mood and acknowledged that change of Commission policy might salvage its flagging reputation.

> [F]ailure of the Federal Government to participate in funding remedial efforts could lead to (1) charges that the Government is indifferent to the health and economic impact of activities conducted in Colorado on behalf of the Federal Government, and (2) litigation and intense public pressure to force the Government to take a positive stance. A coordinated public announcement now of the Government's willingness to assist in correcting the problem by funding and other appropriate means would . . . place the Government in a far more favorable light than such an announcement at some later time when public pressure will have become intense.[40]

Politicians, fearing that indoor radiation would become a campaign issue, pressured the AEC to modify its official policy and take a more active leadership role in solving the problem. The Commission's congressional supporters also feared that continued criticism of the AEC over tailings might further undermine the credibility of the nation's entire atomic energy program at a time when nuclear power was already under attack from environmentally minded Americans. The extent of the tailings problem in Grand Junction, the public furor over it, and the confusion about who was liable for remedial action prompted the JCAE, especially Colorado's Congressman Wayne Aspinall, and senator, Peter Dominick, to hold a congressional hearing on the matter on October 28–29, 1971. The question of remedial action was the first significant issue discussed at the hearing, and the JCAE members carefully reviewed recommendations from state and federal agencies to develop an acceptable threshold for initiating remedial action.[41]

The congressmen and senators at the hearing were particularly interested in a controversial Interagency Steering Committee vote taken in September 1971. During the steering committee's autumn meeting, its medical advisory committee, composed of doctors and researchers from the University of Colorado and from the Mesa County area, had unanimously agreed that "[i]n instances where tailings have been found under and adjacent to buildings in the Grand Junction area, . . . the tailings should be removed and the AEC is the responsible agency for removal of the tailings."[42] Taking that recommendation as its starting point, the medical advisory committee then counseled the steering committee that all tailings within ten feet of any habitable structure should be removed if the gamma radiation readings from the property exceeded 20 microroentgens per hour. This recommendation was based on its findings that such gamma radiation usually demonstrated the presence of tailings under a building and strongly indicated elevated radon levels within it. This curbside gamma radiation test to determine eligibility for remedial action under the Surgeon General's guidelines would have saved time and money. It offered Grand Junction residents an immediate analysis of their property without the usual intrusive, long-term radon investigation. It also meant that hundreds of Grand Junction structures qualified for corrective attention without any direct measurement of radon levels within each individual structure. AEC opponents of the proposal objected to the gamma radiation test as a basis for cleanup and claimed that more testing needed to be completed, including comprehensive indoor radon analysis, before a remedial action program could be undertaken.

After a tense daylong meeting peppered by "sharp dialog," on September 21, the steering committee voted 4–3 to adopt the recommendation of its medical advisers. The state and EPA representatives voted in favor of the recommendation; the AEC and the single remaining USPHS representative voted against it. After the vote, the AEC representatives were "visibly shaken and angry" that the public health experts supported a course of action that they believed was unsound, if not altogether reckless. For days after the vote, officials from the EPA who disagreed with their colleagues' votes, AEC administrators, and JCAE members tried to persuade the steering committee's EPA representatives to change their votes, but with no success. The meeting, after all, was over and the vote was public knowledge. To change after the fact would accomplish little, other than show that the EPA was incapable of making up its mind on environmental matters. More important, however, the vote accentuated for the public the philosophical gulf that still existed between the health experts, who adopted a prudent position toward the tailings, and the AEC, which displayed its typical reluctance to acknowledge uranium tailings health risks.[43]

It is difficult to determine what additional data would have been necessary to change AEC officials' minds at the time about prompt remedial action. Of the 12,351 Grand Junction structures that had undergone preliminary screening, 2,870 contained tailings in part of their construction, and some estimates placed the total number of affected structures as high as 4,000. It had been conclusively established at the time of the September 21 steering committee vote that the presence of tailings under or around a building dramatically increased the likelihood that the structure had elevated gamma radiation and radon levels. It had also been shown that many Grand Junction buildings had sufficiently elevated radon and gamma radiation levels to make them candidates for remedial action according to the Surgeon General's radiation exposure guidelines. Yet, despite this evidence, the AEC voted against taking any corrective action until the entire area could be thoroughly screened, an expensive process that would have required perhaps years of detailed investigation. Under the circumstances, it appeared even more suspicious to the press, concerned politicians, and the public that the AEC should resist prompt remedial action.[44]

Under questioning at the congressional hearing, AEC officials justified their negative steering committee vote on September 21 by explaining that Commission experts viewed a blanket requirement, such as the one proposed to the steering committee, as inconsistent with the Surgeon General's gamma radiation and radon gas exposure guidelines. AEC officials testified that using gamma radiation readings from the property to predict radon gas levels in the homes was too imprecise a method on which to base a far-reaching and expensive cleanup program. They thought that remedial action should not be undertaken without a thorough screening of the homes themselves, a process that would have been time-consuming and expensive. Witnesses noted also that the amount of long-term exposure to radon gas in a home was difficult to predict and that any decision to remove tailings should take into account mitigating factors, including the location of the tailings, ventilation, and occupancy rates. They further testified that the members of the steering committee who had voted in favor of the tailings removal provision had based their decision on insufficient data about how effectively external gamma radiation readings could predict indoor radon gas contamination. The EPA, in particular, supported the AEC position: apparently, by the time of the congressional hearing, EPA officials had come to regret their agency's affirmative steering committee vote.[45]

The Colorado representatives at the hearing expressed their disagreement with the AEC about the need for immediate corrective action. They believed that the AEC's vote against recommending prompt action and its subsequent explanation were further attempts to downplay the indoor radiation problem and delay the remedial process at the expense of the public well-being. They also testified that the state continued to support the medical advisory committee's conclusion that the AEC was the agency primarily responsible for implementing remedial action.[46]

In his introductory remarks, chairman Aspinall had indicated that the hearing was evidentiary and not intended to determine liability or fault for the indoor radiation problem in Grand Junction. Nevertheless, although expert witnesses debated the scientific merits of tailings management, financial liability for resolving the tailings contamination was the major, underlying theme of the contentious hearing. Federal agencies, especially the AEC, feared that federal assistance in resolving the indoor radiation problem in Grand Junction would set an expensive and burdensome precedent for dealing with tailings contamination in other cities. Yet precedent-setting was just what Senator Dominick had in mind. "The difficulty with which we are faced here, as I see it," he said, "is . . . who is going to pay for what. What we are really doing is trying to establish a precedent one way or another which will apply not only to Grand Junction but to every community in the country where there are tailings."[47]

No one at the hearing disputed that a comprehensive remedial program would be expensive. Thousands of homes were affected. As witnesses noted, for example, there were 120 cubic yards of tailings under one home, 106 yards under another, 182 yards under a gasoline station, and 150 yards under a motel. Fifteen of thirty-one schools in Mesa County had tailings under them. An AEC-sponsored engineering study focusing solely on homes and schools estimated that the cost of removing the tailings from each of the 240 homes in the sample study ranged from $43 to $15,000, with an average cost of $3,220 per home. Based on these figures, the cost of removing the tailings from the 4,800 homes probably contaminated by tailings would have been about $15 million. One estimate placed the cost as high as $20 million. Removal of tailings from the schools alone was estimated at nearly $2.8 million. Even conservative estimates for removing tailings from only the homes and schools with the highest and most dangerous radiation levels came to $3 million or higher. Presumably removal of tailings from buildings yet to be identified by the study would add significantly to cleanup costs. The state had good financial incentive, as well as public health reasons, to fix liability on the AEC, and vice versa.[48]

The one federal agency most implicated in the tailings problem was the AEC, but it disavowed any financial responsibility for the Grand Junction situation. AEC officials at the hearing repeated arguments the

Commission first made in the 1950s and early 1960s to respond to the tailings water and air pollution problems. Commission witnesses at the 1971 indoor radiation hearing again denied liability, to begin with, on the grounds that the Climax mill had been an independent contractor. Since the terms of its procurement contract with Climax did not allow the AEC to control access to the mill tailings, the AEC had no authority to assert any such control.[49]

Commission representatives further testified that the AEC could not be held accountable because the agency's legal authority was limited to protecting the public from radiological hazards of uranium or thorium in the ore. First, according to the AEC, it could not protect the public from hazards of uranium and thorium in the tailings because there was less than 0.05 percent of those elements in the tailings. Apparently, even though the AEC had earlier prevented the discharge of radioactive effluent from operating mills into the environment regardless of their uranium or thorium content because it considered those discharges to be an integral part of the milling process, the AEC did not consider the transfer of tailings to contractors for construction to be an equally integral part of the milling operation so as to allow regulation of the tailings to prevent them from being used in buildings This is ironic since the agency was aware at least as early as 1959 that some mills had made such transfers for construction purposes.[50] Second, the AEC claimed, the mandate to protect the public from radiation hazards applied only to guarding citizens from radiological hazards of uranium or thorium; the mandate did not extend to radiological hazards of radium, the chief source of the radon and gamma radiation contamination. Radium had historically been regulated by the states and was specifically excluded from federal control under the Atomic Energy Act and, therefore, the states were responsible for protecting the public from radium exposure.[51]

AEC witnesses further asserted that the AEC was not liable because Colorado had officially assumed responsibility for the tailings. This position was based on a 1968 "Agreement" between Colorado and the AEC in which the AEC transferred some of its regulatory powers to the state, thereby releasing it from regulating certain categories of radioactive materials. It was a strained argument; it was not at all clear just what authority over tailings was transferred to the state by the Agreement. Congress passed an amendment to the Atomic Energy Act in 1959, titled "Cooperation With States," to "promote an orderly regulatory pattern between the Commission and State governments with respect to nuclear development and use and regulation of byproduct, source, and special nuclear materials."[52] A state was permitted to perform some of the Commission's former functions after its governor entered into an "Agreement" with the AEC. However, the Atomic Energy Act, together with Commission regulations, freed the agency from exercising control over "unimportant" fissionable materials, including tailings, which by regulatory definition was material containing less than 0.05 percent by weight uranium or thorium. No federal license was required for handling such "unimportant" materials. Because the laws and regulations limited AEC authority over tailings before the Agreement was signed, it is questionable whether there was any AEC jurisdiction over tailings that could be ceded to the state by the Agreement. Indeed, it was likely that the regulatory authority granted to the state under the Agreement did not grant powers the state did not already have over atomic matters before the 1959 amendment or the 1968 Agreement.[53]

The uncertainty over what powers were transferred to the state by the Agreement was compounded by the ambiguous language of the Agreement itself. The 1968 Agreement between Colorado and the AEC, like all such agreements, generally incorporated the statutory language. Within the Agreement, however, each party acknowledged that the limits on their rights and responsibilities regarding protection against radiation hazards "are not precisely clear."[54] In other words, the Agreement was intended to broaden state authority over atomic energy, but the vagueness of the Agreement language meant that the extent of state control was not conclusively defined in the Agreement. Certainly the nearly ten years of strident debate over which government was obliged to manage the tailings waste could not be resolved by an Agreement that, by its own language, did not specifically clarify federal and state responsibilities.

Furthermore, AEC supremacy in matters of radiological safety was implied by the limitations placed on the state's authority under the Agreement. According to the Atomic Energy Act, before the AEC could transfer any regulatory power to the state, the state governor was required to certify that the state's regulation program was adequate to safeguard the public health, though it was not made clear what the requirements of such a program should include. The Commission had to determine whether the state's regulation program was compatible with its own, but, again, it was unclear how the AEC was to assess the state program. Finally, an agreement could be terminated unilaterally by the Commission for the protection of the public health and safety. The AEC, therefore, remained primarily responsible for protecting the public despite provisions in the agreement surrendering some regulatory authority to the state. Because the AEC retained ultimate authority for shielding the public from radiological hazards, its claim that it was not liable for the indoor radiation problem because it granted away its regulatory authority to the state by Agreement was questionable.[55]

Not surprisingly, Colorado witnesses believed that the AEC, not the state, the community, or the millers, was clearly liable for the cleanup. They had many motives for assigning liability to the federal government. The state could not afford remediation costs that were estimated in the millions of dollars. It also needed to forcefully address any lingering questions that the state was somehow responsible for the Grand Junction problem. State and local authorities were very reluctant to compel property owners to foot the bill to rid their homes of radioactive contamination. Gamma radiation and radon are undetectable without sophisticated equipment, and home owners, therefore, did not know whether their buildings were contaminated. In most instances, the owners did not even know that tailings were under or part of their homes. It would have been equally unfair, state officials believed, to hold contractors entirely liable for the problem because, at the time of construction, they had not been aware that the tailings were unsuitable for building homes, schools, and businesses. Builders had no warning of any problem involving use of the tailings until 1966, when the state and Climax denied them access to the tailings. Indeed, state and federal health officials had encouraged the use of tailings in certain projects, such as road construction, as an ideal method of disposing of the tailings, without giving any admonition that the tailings might be too hazardous for use in private and public buildings. That use of the tailings was encouraged for some building purposes had arguably given contractors the impression that the tailings were suitable for similar subgrade construction purposes, such as under residential and business foundations. Climax representatives noted that their company was not liable because until 1966 they also had been unaware of any problems arising from using tailings in certain types of construction.[56]

The state's hearing witnesses accused the AEC of "extreme irresponsibility" in dealing with the Grand Junction situation. The Atomic Energy Act, to be sure, stated that the AEC had jurisdiction only over wastes that were "byproduct materials," that is, man-made wastes. Radiation sources, including radium, that were unsuitable to produce fissionable material were not subject to AEC regulation and were historically within state jurisdiction. According to the state, however, although radium in its natural location in the soil or the small amount contained in medical apparatuses was clearly within its control, the fact that radium was concentrated in the tailings piles because of mining, transporting, and processing of uranium ore for sale to the AEC negated any federal claim that the tailings were "natural." As a man-made radiation source, the tailings, in the state's view, were subject to federal control and thus any access to them, or transfer or use of them for construction, was and always had been an AEC responsibility.[57]

Witnesses from the Colorado Department of Public Health testified that they could not be held responsible for the tailings, because, until 1966, the state had not even been aware that the tailings were being used for residential construction in Grand Junction and that the AEC had not been regulating their transfer. AEC representatives responded that in 1961 the Commission had officially notified several states and the uranium mills by letter that the tailings contained elevated amounts of radium and might require state regulation. AEC officials even produced an undated file copy of this letter, but the state representatives testified that they never received it. Colorado officials contacted the eight other states that had allegedly received the letter, but none could locate it in their files either. The AEC claimed that even if the states did not receive the letters, on several occasions after 1961 it had informed the state health agencies that they should decide if tailings use should be allowed on a case-by-case basis. State officials maintained that there had been no warnings.[58]

Compounding the confusion regarding the alleged warnings about the dangers of radium in the tailings, the Colorado Department of Public Health and the AEC also disagreed about when the Commission had first learned of the indoor radiation problem. The AEC claimed that the problem did not come to its attention until 1969, well after the contamination occurred in Grand Junction. It was a ludicrous assertion that further undermined AEC credibility. One of the agency's own witnesses testified at the hearing that the Commission had been "informally" notified of the possible problem of indoor radiation in 1966. In 1962 the USPHS, in cooperation with the industry, the states, and the AEC, published a technical report cautioning that indiscriminate use of tailings as construction materials should be prevented because of the threat of radiation exposure. As early as 1960 AEC experts themselves recognized that mill tailings used as landfill and roadbeds in housing developments could increase the public's exposure to radiation. On at least six separate occasions between 1958 and 1960, AEC officials responded to inquiries from millers about the transfer of tailings for construction. In two of these instances, millers specifically expressed their concern about radiation from tailings. The AEC informed the mills that it did not have legal authority over tailings or their transfer.[59]

In light of this evidence, no one believed that the AEC did not know until 1969 of the potential for harm from tailings used in construction of habitable buildings. At the very least, according to Governor Love, the AEC "either knew or should have known they were releasing potentially dangerous materials from their jurisdiction."[60] Roy Cleere, executive director of the Colorado Department of Public Health, summarized his state's position at the hearing:

> Since the advent of the atomic age, everyone has looked to the AEC as the agency with the greatest expertise in radiation matters. Its staff—above all others—should have been able to recognize the long-term hazards of radioactive uranium wastes and provide the protection which the public felt it was receiving. Without the AEC's procurement program, there would have been no uranium tailings problem in Grand Junction. With proper control of the Climax tailings pile, we would not have the situation we have today; namely 2,870 structures confirmed as having radioactive tailings underneath or up against them. This figure is from a total of 12,351 screened so far. Another 4,000 or so locations remain to be screened. . . . Wouldn't you have thought that the Atomic Energy Commission scientists and Climax officials would have recognized the problem before it began?[61]

Glen Keller, president of the Colorado State Board of Health, was considerably more blunt in his assessment of the AEC's role in the matter.

> We have observed the AEC first to deny the possibility of a problem, then to admit possibility but pooh, pooh probability, then to recognize a problem and deny responsibility and finally, in the last few weeks to engage frantically in behind-the-scenes efforts to avoid the recommendations of the Interagency Steering Committee and the Medical Advisory Committee that the tailings be removed. . . . The State of Colorado cannot afford the cost of removal. The people of Colorado cannot afford to leave the tailings in place. This committee cannot afford to be blind to our need.[62]

Despite such impassioned pleas, extensive testimony from the witnesses, and hundreds of pages of supporting documents, the congressional hearing on indoor radiation came to an inconclusive end, leaving the residents of Grand Junction even more uncertain about their future. The key issues of what criteria should be used to determine whether tailings should be removed from under the Grand Junction homes, schools, and businesses and who should pay for the removal remained unanswered.[63]

The AEC's defensive attitude to the Grand Junction situation did little to bolster its already declining credibility regarding environmental issues and its commitment to public safety. Tremendous public and political pressure generated by the hearing, coupled with growing doubts among Commission administrators about the AEC's rigid stance against assuming liability, embarrassed the agency to such an extent that it was forced to modify its long-standing position that it had no liability to decontaminate Grand Junction. While the Commission continued to adhere to its traditional legal stance that it was not empowered to regulate tailings, it rendered that point effectively moot by admitting finally that extensive remedial action in Grand Junction was necessary. At a meeting with Governor Love in December 1971, AEC chairman James Schlesinger candidly admitted that allowing tailings to be used in construction had been a mistake and that Grand Junction contractors, the state, and the AEC had a "moral responsibility" for the resulting contamination. Although the Commission had no legal obligation, Schlesinger stated, it might share the cost of solving the problem. He reaffirmed the new AEC position five months later, stating that the Commission did not believe the Grand Junction home owners should bear the full financial burden of remedial action. He conceded that because the tailings were the result of milling operations that had furnished nuclear materials for national defense and security, the AEC had a "compassionate responsibility" to assist in a remedial action program.[64]

In fall 1971, a few weeks before the controversial congressional hearing, Aspinall had promised his Grand Junction constituents that he would introduce legislation for federal aid to the city if federal responsibility could be demonstrated. Although he remained personally unconvinced that clear responsibility had been shown, Aspinall bowed to the storm of controversy that erupted from the hearing and conceded that federal financial assistance in Grand Junction was necessary. Dominick and Aspinall, with tacit AEC approval, introduced identical bills in the Senate and House, on February 9, 1972, to finance removal of the tailings from under and around contaminated Grand Junction buildings. The principal feature of the bills was that the federal government would pay 75 percent of the costs of a state program to assess and remove the tailings. AEC officials supported the basic objective of the bills and the final legislation but were uncomfortable with early versions of it. They first criticized language in the draft bills suggesting that home owners were being exposed to radiation. They were also concerned that there was no provision requiring the state to pay the remaining 25 percent of costs. The agency feared that, in light of the cost estimates highlighted during the indoor radiation hearing, the original versions of the bills understated the amount necessary to remove the tailings. The Commission was particularly alarmed by proposed language that would require the AEC to take the lead in determining the need for action and to select appropriate remedial strategies, because the AEC wished to continue its policy of making the states primarily responsible for completing all remedial action.[65]

The final version of the bills, which became law on June 16, 1972, incorporated some of the changes proposed by the AEC and allocated $5 million in federal funds for the program. On October 17, 1972, the AEC and Colorado signed an agreement outlining the separate state and AEC responsibilities. The Commission agreed to develop criteria and priorities for action based on the Surgeon General's guidelines, and the state agreed to oversee the performance of the actual work and to deal directly with property owners. The agreement also mandated close cooperation among local, state, and federal agencies to facilitate the cleanup. The law was a Pyrrhic victory for the Commission. The agency's long-standing legal position that it was not primarily responsible for off-site tailings pollution was vindicated by the very fact that Congress passed new legislation to assign responsibility for the Grand Junction tailings problem to the federal government. At the same time, the new law swept aside the agency's traditional interpretation and the federal government assumed primary financial responsibility to remedy the indoor tailings pollution in Grand Junction, something the state had demanded for years.[66]

The issue of indoor radiation contamination from tailings coincided with the public's growing interest in environmental matters. The new environmental movement of the 1960s, focusing as it did on the dangers of pollution to human health, introduced a new urgency to the public debates about tailings hazards. Mill waste was one of many threats, including oil spills, pesticides, and air and water pollution, that seemed to race ahead of their technical and social solutions, overpowering traditional, conservation-style environmental remedies. The Commission's arguments to avoid responsibility for tailings during the radium water pollution and atmospheric contamination problems of the late 1950s and early 1960s, while perhaps justified on legal and regulatory grounds, raised suspicions among health experts and westerners about the AEC's commitment to atomic safety.

By the 1970s those same legal arguments seemed absurd to many Americans—obsolete relics from the nation's atomic infancy. The Commission, which in its early existence exemplified the nation's quest for nuclear independence, seemed out of step with the national consensus. The

AEC's steadfast policy of refusing to assume leadership in solving tailings pollution increasingly made the Commission appear insensitive to nuclear safety and environmental concerns at the very time when those issues were coming into the forefront of national attention. Moreover, the Commission seemed unsympathetic to the problems faced by the people of Grand Junction. "Innocent property owners," reflected two contemporary legal scholars, "should not be forced to absorb the damages which have resulted from a governmental decision to encourage nuclear power, especially when it is not altogether clear that the decision is for the benefit of all concerned."[67] The issue of radioactive tailings used by the construction industry thus attracted far more public attention than did the related issues of radium water pollution or atmospheric radon and dust pollution, which arose before the public's new interest in the environment. More than other instances of tailings pollution, the images of radiation directly contaminating homes, schools, and businesses galvanized national interest in the hazards of mill tailings and goaded Congress and the AEC to solve the problem. The uranium mill towns became tangible symbols of thoughtless atomic development. The crisis dealt a serious blow to the public's already waning confidence in domestic atomic energy and hastened the demise of America's romance with nuclear power.

7

Congress and UMTRCA

IN LATE SUMMER 1972, about a year after the JCAE's Grand Junction indoor radiation contamination hearing, the AEC staff concluded that public interest in the matter of uranium tailings had waned. The remedial action agreement between Colorado and the federal government, pursuant to which over $5 million in federal assistance was committed to clean up the city, appeared to have eased the concerns of Grand Junction residents. The AEC's assessment could not have been more mistaken. Although national public and media interest in tailings had waned, hostility to and resentment of the AEC among many Grand Junction residents still simmered. More significantly, the Commission's wishful thinking underestimated the magnitude of public apprehension about tailings and the level of enmity among western state governments to the AEC's lack of leadership in the matter. Grand Junction was only one of several western communities plagued by tailings contamination. After the successful effort by Colorado officials and congressmen to secure federal funding to decontaminate Grand Junction, residents of other western uranium communities reevaluated their own tailings problems. They hoped to repeat Grand Junction's success and expected a similar settlement to deal with the tailings in their cities. State officials in Utah, in particular, were concerned about tailings at the former Vitro Chemical Company uranium mill, located in a mixed residential-commercial neighborhood of Salt Lake City about two miles from the downtown area and the Utah capital. On a clear day, state legislators could see the tailings pile from the steps of the statehouse. Their effort to persuade the federal government to underwrite a remedial action program for the Vitro site triggered a cascade of political momentum in Congress to address the tailings problem once and for all.¹

The Vitro facility was one of six pioneering uranium reduction mills. Like many early mills, it was originally built to process other materials and was converted to meet the nation's growing demand for uranium. During World War II, the Defense Plant Corporation constructed and operated an alumina production facility at the site as part of the U.S. government's National Defense Program. After the war, the War Assets Administration sold the plant to J. R. Simplot Company, which sold it to the Vitro Chemical Company in early 1951. Vitro modified the facility and began processing uranium ore a few months later. The company made its first shipments of uranium oxide to the AEC in October 1951. That commercial relationship lasted until 1965. During its lifetime, Vitro generated between 1.7 and 2.3 million tons of tailings, or approximately one million cubic yards, that it deposited on its 110-acre mill property. The facility's AEC operating license expired in 1968, and the plant was dismantled two years later. The company removed everything of value from the property but left behind its huge tailings pile.²

Utah health officials had been concerned about Vitro's tailings for years. In 1959 local contractors and the Salt Lake City municipal sewer authority asked the AEC whether they could use the tailings for backfill in construction projects. The Commission permitted Vitro to release tailings to contractors, noting that the builders did not need an AEC license to receive the tailings because they contained too little uranium or thorium to be legally considered source material. At about the same time, Utah health officials expressed far more caution about the disposition of Vitro's tailings, prohibiting the sale of three thousand tons to a contractor intending to use them as fill for a new warehouse. Utah's director of public health was concerned that such a release might set a dangerous precedent. Unlike mills in less urbanized locations, Vitro had limited space for tailings storage, and the director feared that it would seek convenient ways to dispose of its tailings that might result in "uncontrolled exposure to present or future users." Moreover, contrary to the prevailing view of the AEC and most public health authorities, by 1961 the Utah health division was even uneasy about approving the use of tailings for subgrade material for highway construction projects, a method of tailings disposal considered by many to be ideal. Recognizing that highways are sometimes relocated, necessitating the reexcavation of the subbase material, the state public health officials expressed their doubts about finding any sure method of preventing the tailings from developing into a contamination problem in the future.³

Utah's long-standing apprehension about the Vitro tailings prompted state health officials to monitor the tailings closely throughout the 1960s. At the insistence of state officials and the AEC, Vitro tested stabilization and revegetation techniques for its pile during the early 1960s. Vitro officials, however, were running out of room on the mill property and continued to stress their need to remove tailings from the mill site. They favored long-term tailings disposal in highway construction or similar projects. Despite state caution about tailings for construction, a 1967 review of Vitro records showed that about 22,000 tons of tailings had been removed and used as construction fill since the mill opened. Builders probably used most of the material in projects such as road and sewer construction. Yet the state's vigilance paid off; unlike Grand Junction, investigators found few buildings constructed with tailings in the Salt Lake Valley. Ongoing state monitoring efforts and company attempts to stabilize the pile could not prevent conditions at the Vitro site from deteriorating, however, especially after the mill closed.[4]

The issue of the final disposition of the Vitro tailings came to a head in late 1972, when Silvex Corporation, a group of land developers, proposed leveling and covering the tailings to convert the Vitro site into an auto racetrack. The Salt Lake City Suburban Sanitation District also wanted to expand its sanitation facility onto the property. The owner of record of the Vitro property, Zion's Securities Corporation, had correspondence from the AEC from the late 1940s assuring Zion's that the land, together with the tailings, could be returned to unrestricted use after termination of uranium milling activities at the Vitro site. By the 1970s that promise was clearly worthless. Developers brought their problem to the attention of Utah state officials. With the Grand Junction agreement between Colorado and the federal government fresh in his mind, Utah governor Calvin Rampton requested that the AEC contribute funds to the cleanup of the Vitro facility, too. The Commission refused. It instead admonished the state to make every effort to ensure that the site not be used "for any activities involving construction or ingress by humans, regardless of the amount of cover placed over the tailings." The AEC also reiterated its traditional legal position that although development of land used to store tailings was absolutely prohibited at facilities licensed by the AEC, the Commission assumed no responsibility for tailings at closed mills for which the operating licenses had expired, including Vitro.[5]

Utah authorities came under intensified pressure in early 1973 from other real estate developers wanting to build on the Vitro property, a parcel of real estate that was becoming increasingly valuable as the population of

Salt Lake City grew and the city expanded. The Vitro pile, however, was such a hazard that the property could not be commercially developed without extensive remedial efforts. The need to salvage the site, the AEC's own recommendation that access to the site be restricted, and the long-standing Utah health department concern over the Vitro facility compelled Governor Rampton to investigate options to clean the property for development. Robert Pendleton, one of the governor's leading scientific advisers, informed him that there was simply no adequate solution available to the state. Pendleton cautioned that the ideal way to render the Vitro land safe was to remove all the tailings to an uninhabited location for stabilization, a huge operation that he believed was beyond the state's abilities. Absent such a "heroic waste removal operation," he could only suggest that the tailings be covered in place and the property fenced off from human access until the radium decayed to an innocuous level, a solution he considered to be wholly impractical since the property would have had to be isolated for over twelve thousand years! Undeterred, developers pressed the state to approve their building plans under whatever reasonable restrictions the health department believed were necessary.[6]

In addition to the problem of tailings on the former Vitro property itself, tailings contamination appeared off-site as well. The long-standing vigilance of the Utah authorities about tailings meant that the Grand Junction experience was not repeated in Salt Lake City. Nevertheless, with the Colorado situation fresh in its mind, in 1972 the EPA conducted a gamma radiation survey in Salt Lake County to screen the area for possible tailings contamination. The initial survey identified 71 radiation anomalies, and closer investigation revealed 17 vicinity locations in which tailings had been used for construction—5 residences, 7 businesses, and 5 vacant lots. At one of the buildings adjacent to the site, the EPA found radon levels as high as 0.3 to 0.5 WL based on the average exposures to the public in those structures. At the time, the exposure limit for uranium miners was about 0.3 WL, based on a forty-hour workweek. For the general public, exposure was limited to about 10 percent of that radiation exposure, or 0.03 WL. In cases in which exposure was near twenty-four hours per day, the recommended limit was 0.01 WL. Ultimately, the Department of Energy identified only 118 vicinity properties that were sufficiently contaminated to warrant remedial action. A few of those properties were contaminated with tailings used for construction. Many, however, were contaminated by windblown tailings. The disclosure of the tailings contamination, so hauntingly familiar to the problem in Grand Junction, understandably caused considerable public alarm among Salt Lake Valley residents.[7]

State officials realized by fall 1973 that they must make some final judgment about the disposition of the Vitro site. The state health division, having already recommended that the only prudent course of action was to remove the pile completely, now went one step further. It recommended that there be no new construction at all within one-half mile of the property until scientific studies demonstrated the most practical method of rendering the site safe for human occupation. Governor Rampton was prepared to provide state money to remove the tailings but only if the federal government contributed, as it had in Grand Junction, the lion's share of the cost.[8]

Faced, on the one hand, with the need to render the Vitro site safe and, on the other, with federal intransigence in paying for remedial action, Utah Senator Frank Moss and Representative Wayne Owens introduced identical bills in Congress to require the federal government to provide 75 percent of the cost of decontaminating the Vitro site.[9] In both language and intent, the bills mirrored the Grand Junction remedial legislation. The JCAE conducted a hearing on the bills on March 12, 1974. The Vitro hearing was far more amicable than the contentious 1971 JCAE hearing on the tailings situation in Grand Junction. There was, for example, much less disagreement among the witnesses at the Vitro hearing about the environmental and health hazards of the Salt Lake City pile and the need to do something about them. The substantive issue, as with the Grand Junction hearing, was liability for cleanup. The state witnesses argued that since the Vitro plant had furnished uranium for the AEC, the federal government had a responsibility to contribute financially to a remedial program.[10] In his prepared statement, however, Governor Rampton framed the issue differently. For Rampton, assessing responsibility for the Vitro tailings was more than a question of liability; it was one of equity.

> In perspective, the Vitro Tailings pile represents an environmental problem which originated within a federally sponsored program.... The federal government in sponsoring this raw materials program acted for and in behalf of the American people from whom it derives its purchasing power. The monetary costs were shared by the total populace and all of us also shared the intended benefits from the program. As our investigations now show us, however, the environmental risks have not been equally distributed, and those in the Salt Lake Valley are experiencing a disproportionate share of the risks, especially the people who live and work in the immediate vicinity of the

pile. The Vitro pile is the only abandoned uranium tailings site located in the center of a large metropolitan area![11]

The AEC representative at the hearing, James Liverman, disagreed with Utah witnesses who claimed that the tailings from the Vitro site needed to be removed. Balancing the costs in human health if the tailings were left in place against the financial burden of removing the tailings, Liverman recommended that the tailings be left in place and isolated from human use almost in perpetuity. Realizing the impossibility of that option, however, he acknowledged that the site, if covered by twenty feet of soil to reduce the amount of radon escaping into the atmosphere, could be used for a parking lot, park, or other similar site in which human occupation was limited. It was a solution that few Utah officials or Salt Lake developers found attractive.[12]

AEC officials recommended that removal costs of the Vitro tailings were unjustified. Yet after their experience with Grand Junction, they themselves were no longer convinced that theirs was the best approach to the tailings problem. Seven months before the Vitro hearing, in a letter to eight western governors, the AEC largely conceded the states' position that the tailings were a greater potential health problem than the agency had historically admitted and asked the states for assistance in studying the most appropriate corrective action. Liverman testified at the Vitro hearing, "The presence of naturally occurring radium in uranium mill tailings is the principal source of public health concern because of the radioactive gaseous radon which continuously emanates. Since radium has a half-life of 1600 years, the problem is a persistent one."[13]

Liverman, joined by W. D. Rowe of the EPA, used the hearing forum to acknowledge a new federal attitude toward abandoned uranium mill tailings. During the 1970s, AEC, later NRC, licensing authority had been used to gradually graft new requirements onto mill company licenses to ensure tailings were discarded in an environmentally sound manner. That policy was greatly expanded in 1970 after the passage of the National Environmental Policy Act (NEPA).[14] Consequently, industry and atomic technocrats believed that the tailings problem at operating mills had been largely resolved. At the hearing, then, Liverman underscored the fact that there were many abandoned mill tailings sites across the nation that fell outside this evolving regulatory scheme and that demanded some form of similar attention. He reminded the JCAE members that the Grand Junction legislation, intended to clean up homes but not to stabilize or remove the tailings pile itself, was not a clear precedent for authorizing federal money to remove and dispose of the tailings at the abandoned and inactive sites. Rather than address the tailings problems at the various locations on a piecemeal, case-by-case basis, the AEC and the EPA suggested that the federal government approach the abandoned tailings "in totality." It was a breathtaking break from the historical federal response to tailings management.[15]

To assess appropriate comprehensive remedial action for all abandoned tailings piles, AEC and EPA witnesses proposed a two-phase study. Phase I involved a preliminary survey of all abandoned tailings sites to determine each pile's condition, the need for corrective action, ownership, proximity to population centers, and the prospect of increased population near the sites. That study would then form the basis of an in-depth Phase II engineering report detailing alternative remedial action measures for each site and estimated costs for cleanup. The EPA believed that legislation would eventually become necessary to return the abandoned mill tailings sites to conditions that would allow unrestricted public use. At the close of the Vitro hearing, the JCAE accepted the broad recommendation of the EPA and the AEC to study the abandoned tailings problem further under Phase I, before coping with the financial and logistical nightmare of attacking all of the nation's abandoned tailings sites.[16]

Senator Moss and the Utah delegation were not happy with the JCAE's decision to undertake further study. From their perspective, the Vitro pile was an existing problem that had already been evaluated fully by both state and federal agencies. The impatience of Utah officials was perhaps justified. By late 1973 the EPA had completed its survey of the Vitro pile and concluded in its report that "tailings be removed if at all feasible and properly disposed in a properly designated location." Paul Smith, an EPA official from the Grand Junction regional office, informed the Utah State Advisory Council on Science and Technology Subcommittee on Tailings "that no further monitoring is necessary to substantiate that the tailings do constitute a clear health hazard and that the existing data conclusively demonstrated that the radon levels exceeded existing health standards." Frank McGinley of the AEC's Grand Junction office and two other EPA representatives present at the subcommittee meeting "strongly concurred" with Smith's views.[17]

Under the circumstances, then, Utah officials did not believe the Vitro problem called for a "package deal." Cost estimates for remedial action at Vitro also continued to soar with each passing year, as did the estimates of how best to isolate the tailings from the public. Keith Schiager, of Colorado State University, concluded that approximately two feet of soil would reduce gamma radiation at the Vitro site to nearly background levels, but reducing radon emissions would require up to thirty feet of soil, not Liverman's twenty feet, or eight solid feet of concrete. Contrary to AEC testimony at the Vitro hearing, Schiager believed that the cost of leaving the tailings in place far exceeded the cost of moving them to a safe location. Utah officials pointed out that the EPA, the Utah Department of Health, and even experts in the AEC and the new Energy Research Development Administration (ERDA)[18] were convinced that removal of the tailings was the only safe and economical solution. The state believed that immediate federal funding was not only necessary but also more than justified under the Grand Junction precedent. Moss was outraged by the delay resulting from the JCAE's decision, claiming that any pause in acting on the Vitro problem amounted to "placing the importance of bureaucratic red tape ahead of concern for the almost infinite hazards of atomic waste."[19]

While Utah officials fumed, the federal agencies immediately began Phase I of the tailings evaluation study. Work proceeded quickly over the summer of 1974, and by October the Phase I report was complete. Ironically, the speed with which it was prepared was a testament to the large body of data, gathered over twenty years, that already existed about tailings contamination. Federal and state scientists and engineers studied the twenty-one inactive mill sites, containing a total of 25.2 million tons of abandoned tailings. Their findings were alarming but merely confirmed what public health experts and environmentalists had known for years. Nearly everywhere improved tailings stabilization and general decontamination of mill sites and the areas surrounding the tailings piles were required. At seven of the locations, containing just under one-half of the total tailings, investigators recommended complete removal of the tailings to more geologically stable and less populated areas.[20]

Phase I of the study concluded with a recommendation that Phase II begin as soon as possible on all of the sites evaluated in Phase I. The ERDA and the EPA estimated that detailed engineering reports on the abandoned tailings sites would take about two years and cost an estimated $1 million, funded primarily by the EPA. Recognizing the unique and extremely hazardous condition of the Vitro pile and the dissatisfaction of Utah state officials with federal handling of the problem, the EPA assured the state that the Vitro site would receive top priority in the Phase II assessment study.[21]

In January 1978, nearly four years after the close of the Vitro hearing, the DOE submitted to Congress the Phase II engineering assessments on twenty-two abandoned tailings sites. Not everyone was pleased with the conclusions. New Mexico, in particular, was concerned that the Phase II report on the abandoned portion of the Ambrosia Lake site (2.68 million tons, 91 acres) fell short of identifying acceptable options for remedial action. Duane Rennels, of the mining and real estate company United International Corp., highlighted several factual errors in the Phase II report on the abandoned pile in Gunnison, Colorado. Utah officials were concerned that the report failed to incorporate all available data on radiological effects, provide sufficient supporting data for its procedures or conclusions, and offer satisfactory long-term mitigation policy options. Although states disputed some aspects of the assessments, the general findings of these reports reiterated the Phase I conclusion that none of the sites had been adequately stabilized for long-term tailings storage. Furthermore, in many cases wind and water erosion of the tailings had spread the tailings beyond the boundaries of the properties, perhaps doubling the risk of cancer in persons living within one-half mile of the tailings. Using the data generated by the Phase II study, the General Accounting Office estimated that leaving the tailings piles uncovered could lead to 339 excess cancer cases over a one-hundred-year period. The Phase II study estimated that remedial action could cost as much as $132.5 million.[22]

After completion of the Phase I study but before work had been finished on Phase II, a series of public disclosures of tailings hazards refocused national public attention on the tailings issue and underscored the need to complete the work. In response to a request from the New Mexico Environmental Improvement Agency, the EPA conducted tests in 1975 on water supplies in the Grants Mineral Belt uranium mining region in northwestern New Mexico. A copy of the draft report, made public by the Public Interest Research Group, indicated that radioactivity levels in water supplies near uranium mines and mills were "intolerable." It also asserted that discharges by Kerr-McGee and United Nuclear Corporation–Homestake Partners violated their license limits, as well as New Mexico and Nuclear Regulatory Commission water quality standards. The people most directly affected by the contamination were the mill employees and their families. The following year, the EPA prepared a report on the levels of radiation exposure to people living near uranium mill tailings in eight western states. The report indicated that individuals could be exposed to "significant" amounts of gamma radiation generated by radionuclides within the piles or from radon and its daughter products exhaled by the tailings. The agency found that the annual dose rate to people living fifty yards from the piles could be as high as sixteen times the NRC standards for annual exposure to the general public.[23]

Walter H. Jordan, a former director of the Oak Ridge National

Laboratory, provided one of the most alarming public disclosures about uranium mill tailings. He pointed out what he believed to be a "gross error" in the NRC's estimates of the environmental health consequences of the uranium fuel cycle, claiming that the "correct" estimate for risk from radon gas was 100,000 times greater than the NRC figure. "Deaths in future generations due to cancer and genetic effects resulting from the radon from the uranium required to fuel a single reactor for one year," Jordan concluded, "can run into the hundreds." When Jordan's findings reached the press in November 1977 they caused a sensation.[24]

These public disclosures about the hazards of uranium mills, coupled with the broad scope and dire conclusions of Phase I and Phase II, reinforced the public's worst fears about the hazards of mill tailings and prompted a flurry of congressional activity. Understandably recognizing that many Americans would find hundreds of "excess deaths" an unacceptably high price to pay for leaving the tailings unattended, congressional representatives introduced four separate bills in the House to address the tailings issue, three intended to solve the problem of existing tailings piles and one to prevent similar tailings problems in the future. Although the three remedial action bills differed somewhat in their specifics, all proposed that the tailings be effectively managed for long-term environmental safety and that the federal government pay the majority of the cleanup costs.[25] These bills constituted the most potent challenges so far to the power and discretion of the federal government in matters of atomic energy.

The Subcommittee on Energy and the Environment of the House Committee on Interior and Insular Affairs conducted the Uranium Mill Tailings Control Hearing in June and July 1978 to solicit comment on the bills. Faced with overwhelming evidence on the hazards posed by uranium tailings, no witnesses seriously contested the need for prompt action. The witnesses disagreed, however, about financial liability for the tailings problem. The chairman of the NRC reiterated the AEC's traditional viewpoint that neither the Commission before it was abolished nor the new Department of Energy and NRC had any regulatory authority over mill tailings after mill operations ended.[26] Without a specific mandate from Congress, the federal government would not pay for reclaiming the abandoned tailings sites.

To overcome this problem, the bills' supporters regularly characterized the tailings problem as a matter of equity. One Utah witness noted that the uranium produced at the mills benefited the entire nation but placed a terrible environmental and health burden on the states and local residents. The Grand Junction case, moreover, was precedent for the notion that justice demanded the federal government assume leadership for corrective measures. A General Accounting Office official testified that the federal government had a "strong moral responsibility" to undertake tailings cleanup and, in any event, was the only entity capable of financing the sort of comprehensive cleanup program envisioned by Phase II. Reflecting on the equity issue with grim humor, the director of the Utah Division of Health suggested that federal action might have been forthcoming sooner had tailings been stored in the central desk drawer of every congressman.[27]

The issue of equitable distribution of cleanup costs between the federal government and the states generated heated debate at the hearing. Two of the bills required the federal government to pay at least 75 percent of the costs, and the third bill proposed that the federal government pay the total cleanup costs. The Carter administration and the DOE believed that the states should pay part of the cleanup bill because they would benefit the most from reclamation. The disagreement over what portion of the cost the federal government should contribute for remedial action focused in part on the federal government's claim that states or private industry might receive financial windfalls at its expense. The DOE maintained that since the states had already received financial benefit in the form of taxes and payrolls from the uranium mills while they were operating, it was fair that they contribute to the cleanup costs. Utah bristled at the suggestion, noting that despite tax revenues for the state, the Vitro mill indisputably had "substantial negative benefit [for the state] . . . and [would] continue to have that negative benefit on into the future." DOE officials, however, had another motive for insisting on state participation. They believed that the states must contribute financing to a remedial program to encourage them to select the most cost-effective solution to the problem, not just the "gold-plated" one. They also felt that states would be more willing to actively participate with the federal agencies in designing and implementing programs if they had a financial stake in them.[28]

Understandably, the huge cost of remedial action for the abandoned tailings piles outlined by Phase II worried cost-conscious representatives. Another issue explored by the House, consequently, was the possibility of profitably recycling the tailings for the residual uranium to recoup some of the costs. Some tailings, particularly those located in Gunnison, were suitable for such reprocessing. Duane Rennels, of United International Corp., testified that his company, after an initial $3 million investment, could recover 65 percent of the residual uranium with a total market value of $9 million. Reprocessing alone, however, did not appear to be a viable option for managing the tailings. Although the reprocessing would likely reduce some of the environmental hazards of the tailings by reducing their acidity and removing some of the residual uranium, it would not remove the radium, the component of the tailings that caused the greatest radiological contamination. Furthermore, the reprocessing would not result in stabilized piles, and, because the radium remained after reprocessing, the tailings would still be dangerous. Creating a final resting place for the Gunnison tailings once they had been reprocessed, Rennels told the committee, would cost between $16 million and $22 million, far exceeding his company's estimated $6 million profit. Rennels's testimony clearly established that reprocessing was not a viable option for tailings management. As was so often true in the history of the uranium industry, producing uranium in the United States would be profitable only with substantial federal assistance. Permanent in-place stabilization or removal remained the only effective way to safeguard the public from tailings.[29]

Although much of the testimony throughout the Uranium Mill Tailings Control Hearing dealt with the problem of remedial action for abandoned tailings and off-site contamination, senators, representatives, and witnesses were equally concerned about the need to enact sweeping legislation to regulate tailings piles from operating mills and to require long-term oversight to prevent the tailings problem from reoccurring in the future. These tailings, after all, represented a significant amount of potentially hazardous tailings. Their concerns were well founded. By the time of the hearing, several states had licensed uranium producers pursuant to the Agreement program created by the 1959 amendment to the Atomic Energy Act of 1954. The amendment delegated to any state that wished to participate in the program the federal government's authority to license uranium processing activities. As of 1978, twenty-five states had assumed regulatory responsibilities under the act and exercised regulatory control over operating mills. The arrangements, however, were not always adequate to maintain control over operating mills. The Agreement States, as they were called, were required to maintain regulatory programs compatible with those of the NRC, but the states' regulations sometimes were less stringent than federal standards.[30] The standards varied from state to state. There were, as well, inconsistencies between state and federal environmental quality standards for both abandoned and operating mills. Moreover, the federal government lacked the authority to compel the states to enforce the higher or more consistent environmental quality standards.[31]

Faced with these seemingly insurmountable problems, congressional supporters worked to reconcile the differences in the various bills but were pessimistic about final congressional action. Congress intended to adjourn in fall 1978, but in mid-August the bills were still bogged down in committee, with no further action scheduled and no action possible until after the three-week Labor Day recess. Prospects for the passage of a comprehensive uranium tailings law appeared dim in 1978. Much to their own surprise, however, supporters overcame conventional wisdom and managed to get the Uranium Mill Tailings Radiation Control Act passed on November 8, 1978, during the hectic final hours of the Ninety-fifth Congress.[32] Considering the obstacles, that it was passed at all was a measure of how much support the idea of mill tailings control had in Congress by the late 1970s and the extent to which environmental protection consciousness had diffused throughout the nation. In passing UMTRCA, Congress completed the process of undermining the traditional federal position to tailings it had begun a few years earlier with the Grand Junction remedial legislation. After more than two decades of federal equivocation, this amendment to the Atomic Energy Act of 1954 finally swept aside the vestiges of the legal and regulatory foundation on which the AEC and the DOE had traditionally declined accountability for the tailings. The law acknowledged federal responsibility for remedial measures, assigned primary liability for uranium mill tailings cleanup to the federal government, required decisive federal leadership in resolving the problem, and dispelled any remaining doubts that the public health and environmental hazards from tailings were serious enough to warrant cleanup and federal regulatory control. UMTRCA was also the first time that Congress, or any federal agency, addressed the environmental consequences of milling with a long-term perspective—in effect, finally validating opinions expressed by public health experts for nearly two decades. Underscoring Congress's recognition of the need for federal leadership, the first sentence of UMTRCA stated:

> The Congress finds that uranium mill tailings located at active and inactive mill operations may pose a potential and significant radiation health hazard to the public, and that the protection of the public health, safety, and welfare and the regulation of interstate commerce require that every reasonable effort be made to provide for the stabilization, disposal, and control in a safe and environmentally sound manner of such tailings in order to prevent or minimize other environmental hazards from such tailings.[33]

Title I of UMTRCA established a "program to assess the tailings at inactive sites and to provide remedial action at such sites, including the reprocessing, as appropriate, of tailings . . . in order to stabilize and control such tailings."[34] It directed the DOE to complete remedial action on twenty-two[35] inactive uranium tailings sites at which all or a substantial portion of uranium was processed for sale to a federal agency and which no longer had a license to process uranium as of January 1, 1978. It also mandated remedial action for nearly 5,200 contaminated vicinity properties. This section of the act stressed federal, interstate, and state cooperation and cooperation with Native American tribes and persons who owned or controlled inactive tailings. Perhaps recognizing the historical atmosphere of suspicion that had existed between the AEC and state and federal public health agencies, Congress in this title also mandated interagency cooperation and coordination among the DOE, the NRC, and the EPA in enforcing the law, with the states and tribes playing active advisory roles. Title I directed the DOE to pay 90 percent of the actual costs of any remedial action at designated sites, including the costs of acquiring a tailings site or a disposal location. The law required states to pay the remainder of the costs from any nonfederal funds. No federal assistance was to be available for active uranium processing facilities, a provision reflecting the congressional intent that costs of protection against hazards be borne by the commercial uranium processing industry whenever possible.[36]

Congress clearly intended that its participation in reclamation of abandoned or inactive tailings piles not be seen as a precedent for future federally financed cleanups. It emphasized that abandoned tailings were a unique case. These unlicensed tailings sites had escaped federal control because the dangers of tailings were not clearly understood when the uranium industry was booming. Such was not the case at operating mills, because those sites had been subject to NRC regulation after 1970 as a result of NEPA.[37] Therefore, Title II of the law made millers responsible for cleaning up all the tailings stored at their facilities, including the millions of tons produced under government contract before 1970.

Title II of the act was designed to set health and environmental standards for mill tailings to achieve "permanent isolation of tailings and associated contaminants" without ongoing maintenance.[38] It directed the NRC to regulate uranium mill tailings at those processing sites having an active license as of January 1, 1978. Perhaps the more important of the UMTRCA sections, it ended any equivocation about the federal government's authority over the waste disposal practices of the uranium milling industry. It strengthened the federal and Agreement States regulatory scheme over active mill operations, regulating uranium mill tailings at operating mills in a manner designed to eliminate the need for future remedial actions. The financial surety program of this section forced millers to post bonds to ensure performance of approved decommissioning and reclamation plans. Legislators believed that NRC licensing requirements and NEPA had solved the tailings problems at active mills, but Title II reinforced the authority of the NRC to regulate the production and disposal of mill tailings at active mill sites. Under Title II, the EPA was required to develop general standards for tailings reclamation at operating mills, after which the NRC would write specific rules and apply them to individual mills through the process of issuing and amending licenses. It also ensured the application of minimum federal standards to such regulation activities in Agreement States in order to "stabilize and control such tailings in a safe and environmentally sound manner and to minimize or eliminate radiation health hazards to the public."[39]

Under the terms of UMTRCA, responsibility for managing the abandoned tailings fell to several agencies. It was a daunting assignment; about 39 million cubic yards of uranium mill tailings and other hazardous constituents, such as abandoned buildings and equipment, covered about 3,900 acres, and thirteen of the sites required removal of the tailings to off-site disposal facilities. To comply with the law, the DOE created the Uranium Mill Tailings Remedial Action Project (the UMTRA Project) to perform surface remedial action at inactive tailings locations and off-site contaminated properties in ten states and on tribal lands.[40] A subsequent amendment to UMTRCA recognized the threat to groundwater supplies and authorized subsurface remediation.[41] In both the surface and subsurface programs, the DOE was designated as the lead agency, and as such responsible for the overall management of the UMTRA Project. The law assigned the NRC federal regulatory oversight of the project. In general, the project had to comply with NRC regulations,[42] and the NRC provided technical and regulatory review for project documents, such as remedial action plans, completion reports, and certification reports. NRC licensed completed disposal sites for long-term care. The project also had to comply with EPA regulations,[43] and the EPA was specifically charged with establishing health and safety standards for disposal of the hazardous material. The DOE entered into cooperative agreements with several states, the Navajo Nation, and the Hopi Tribe to undertake both surface and groundwater remediation programs. The individual states were expected to participate fully in the UMTRA Project, as well as pay 10 percent of the costs. Remedial action plans for Indian lands were completed in consultation with the Bureau of Indian Affairs, and Indians did not have to pay the 10 percent contribution demanded from the states. All remedial action recommendations were subject to the National Environmental Policy Act,[44] which required any federal agency to prepare a detailed statement that identified and analyzed the environmental impacts of a proposed course of action that may affect the quality of the human environment.

UMTRCA was not a perfect solution to the tailings problem. The act strengthened the regulatory framework for managing mill tailings, but it did not address similar hazardous wastes generated by uranium mining.[45] It also overlooked the problem of active mills under federal license that were not covered by the abandoned tailings remedial action provision of Title I but which could not meet the active mills tailings management requirements of Title II without suffering significant economic losses. UMTRCA likewise did not adequately resolve the problem of "commingled waste," that is, tailings produced as a consequence of both private and public contracts. Implementing these requirements of the legislation without seriously burdening the already declining uranium industry was impossible.[46] UMTRCA also empowered federal agencies to encroach on traditional state autonomy in matters associated with mineral production. Experience demonstrated that national interests regarding atomic energy, such as uranium production, usually took precedence over state and local interests in public health and environmental protection and that state governments were usually more responsive to the concerns of local residents. Several states, such as Colorado, had taken the lead in tailings management and were already diligently working to mitigate tailings pollution. Yet, to deal with the problem in a comprehensive manner and eliminate the patchwork of individual state tailings regulations, for all practical purposes UMTRCA eliminated state autonomy in tailings matters that existed prior to the act by mandating that all state regulations conform to federal standards for tailings management.[47] Perhaps most significant, there was little assurance that federal agencies involved in the development and enforcement processes would aggressively carry out their duties under the act. Indeed, environmentalists worried that the deeply rooted habits of the AEC, which they feared had carried over into the NRC and the DOE after the AEC was dismantled, militated against assertive federal tailings management. Their fears were confirmed by the bureaucratic inactivity and outright hostility that were hallmarks of environmental policy during the Reagan administration of the 1980s.

Despite its shortcomings, UMTRCA was a watershed in the history of the uranium milling industry, and in the public policy regarding atomic energy as a whole. On balance, there can be little doubt that UMTRCA was a significant improvement in the management and regulation of mill tailings. It marked the first time that the federal government officially acknowledged that uranium mill tailings pollution, produced as a consequence of its uranium acquisition program, required substantial federal assistance to solve—and to prevent in the future. UMTRCA clearly signaled congressional impatience with traditional policies of legal and bureaucratic resistance that were relics of an earlier atomic era. It settled legally the ongoing debate between public health experts and atomic technocrats about whether the mill tailings were hazardous. UMTRCA helped to end, at least officially, the federal government's practice of assuming that health and environmental concerns were less important than nuclear energy independence. The act also ended the pattern by which state governments, federal health agencies, and congressional representatives were required to confront, on a piecemeal basis, the federal government's authority in matters of tailings pollution. UMTRCA officially marked the end of the era of federal government resistance to the issue of mill tailings regulation and signaled that the federal government had finally recognized that it had a legal and moral obligation to solve the mill tailings problem.

Above all, UMTRCA represented a fundamental shift in the way the nation approached atomic energy. For more than two decades, the federal government pursued a relentless campaign to convince Americans about the benefits of atomic power. UMTRCA demonstrated official acknowledgment that atomic energy came with high environmental costs, and those costs must ultimately be borne by the federal government, which was instrumental in creating the problem in the first place. For better or worse, from the time of UMTRCA forward, the federal government would be forced to justify its atomic energy claims to a public that had grown cynical and suspicious of blanket declarations about the benefits of the atom.

8

Closing the Circle

AMBROSIA LAKE, NEW MEXICO, is haunted by ghosts. For thirty years, its mills processed uranium ore from mines scattered across the arid, prehistoric lakebed and surrounding area a few miles from the town of Grants. During its heyday in the 1950s and early 1960s, the area was alive with activity. The sound of signal bells directing the hoist towers above the mine shafts cut through the desert air. The smell of hot oil, heavy machinery, diesel fuel, and mill chemicals mixed with the sage of the desert, especially after infrequent rain showers, to render a heady industrial perfume. Yellowcake poured from the mills. Lights burning over the mines and mills created a perpetual daytime. Topside and underground, the scene at Ambrosia Lake was one of organized turmoil and purposeful chaos.

The bounty did not last. Uranium mining and milling became a risky business during the late 1960s. In retrospect, the signs of the difficult transition from a government-supported uranium industry to a wholly private one appeared early. Few at the time, however, were prescient enough to understand that the uranium industry was living on borrowed time, surviving only with massive federal assistance and permissive environmental standards. In May 1956 the AEC announced the details of its new domestic procurement program for the period April 1, 1962, through December 31, 1966. That new policy continued government support to "maintain a high rate of exploration and development" but signaled that the AEC expected the private atomic power industry to shoulder a greater financial burden. Yet only a little over a year later, in late 1957, new uranium discoveries and expanding milling capacity led Jesse C. Johnson, AEC director of raw materials, to announce that "it is no longer in the interest of the Government to expand production of uranium concentrate." The surging uranium production threatened to escalate federal uranium purchasing costs. In 1958 Congress refused to continue funding the unlimited uranium acquisition program, and on November 24 the AEC announced that it would continue to subsidize, between 1962 and 1966, only those companies that had government purchase contracts before November 11, 1958. In an effort to reduce its financial commitments to uranium producers, cushion the blow of falling uranium prices for the mining and milling companies, and jump-start the moribund atomic power industry, the AEC removed legal impediments to private sales of uranium concentrates in 1958. Despite the cutbacks, by 1962 uranium production still exceeded government demand. It was equally clear that private uranium sales to power companies could not support the domestic uranium industry—in fact, no private uranium sales were made until 1966. Consequently, the AEC announced in late 1962 yet another change in policy to wean the industry from its historical dependence on the federal government. The Commission agreed to "stretch out" its purchase commitments again, this time extending its support from 1966 through 1970. After that, the uranium industry was on its own. The AEC also shielded domestic producers from Canadian competition and in 1966 placed a total embargo on all foreign uranium.[1]

During the late 1960s and 1970s, uranium prices fluctuated as the industry tried to survive without extensive federal price supports. It proved to be a futile struggle. Despite short-lived rallies, the uranium industry was on a steady downward slide. The nation was awash in cheap uranium. Competition in the international market, excess world uranium production, and declining domestic demand undercut American producers. Cancellation of new atomic energy power plants after the 1979 accident at Three Mile Island sealed the industry's fate. Mines and mills closed, one after another, as demand and prices for uranium sank. Eventually, even the giants fell. In 1985 Kerr-McGee closed its last remaining uranium property and dissolved its uranium subsidiary, Quivira Mining Company. At its peak Kerr-McGee alone employed 2,300 workers. In 1989 Homestake announced the shutdown of its Ambrosia Lake uranium operations. Only a month earlier, Chevron Resources had announced that it was closing its mine, the last in the area. Local miners and mill employees, many of whom had lived and worked in Grants for their whole careers, pondered their futures without uranium.[2]

Today, Ambrosia Lake and Grants stand as stark reminders of the collapse of the uranium industry. There are fewer people, and the national retail and restaurant chains are gone. Santa Fe Avenue, the main street through town, is littered with vacant buildings. The saloons, which once throbbed with energy on payday, are shuttered. On the outskirts of town, a salvage yard displays the skeletal remains of a dead industry—rusted hulks of overturned uranium ore cars, head frame pullies, and heavy machinery sitting in pools of clotted oil. The highway leading to the mining district, once packed with traffic of mill employees and miners commuting to work in round-the-clock shifts, is usually empty. Bunches of tough desert grasses grow in pavement cracks. The air is cleaner, and the chemical smell of the mills is gone. On most days, the only sound is the wind. The abandoned mines are flooded and the head frames are gone. Funnels that once drew fresh air into the underground shafts dot the buff-colored horizon like giant, atomic-era mushrooms. When the sun sets, the only light comes from the moon and the stars.

The most startling feature in the desert landscape is the geometrical mound of black basalt rock that entombs the 2.6 million tons of mill waste from the decommissioned Phillips Petroleum mill, which operated from 1957 through 1962. Fences discourage close inspection of the disposal cell, and warning signs remind its infrequent human visitors that the edifice is a radioactive waste site. Similar tailings disposal cell sites have become recent additions to the landscape of many western states.

The enactment of the Grand Junction remedial legislation, and later UMTRCA, calmed the citizens in the uranium towns and cities. After three decades of ineffective federal leadership, they believed that the tailings problem would finally be resolved. Unfortunately, what Congress declared should happen and how it was carried out in practice were two very different things. In Grand Junction, for example, the problem of indoor radiation contamination continued well after Congress approved the remedial legislation in 1972. The Grand Junction project was plagued by so many difficulties that it did not accomplish the original legislative intent in an effective or economical manner. One shortcoming of the project was the uncertainty about how to pay for the cleanup of tailings that were not located under or near structures and therefore did not come under the purview of the legislation. The Colorado Department of Public Health had identified more than 2,300 of these sites by 1978, many of which, such as vacant lots, were otherwise suitable for development. The state and federal government had no obligation to pay for the cost of cleaning up such tailings. Another problem was that the initial appropriation was far too small to cover the total cleanup bill. Originally, the law had provided for $6.7 million in cooperative state and federal funding, and further congressional action increased the federal funding to $9.5 million, for a total of $12 million after the state's contribution. Financial estimates made in 1978, however, showed that completing the remedial action would probably cost closer to $14 million, perhaps more. In fact, the final cost would ultimately eclipse even that.[3]

The most significant obstacle delaying remedial action in Grand Junction was the manner in which it was carried out. Because contractors bid for jobs on each individual location, rather than on the project as a whole, project managers found it difficult to attract construction contractors interested and qualified to undertake the work, resulting in further project delays. Owners were not required to clean up their properties or to seek help through the program. A few home owners, weary of promises, suspicious of government action, or simply unwilling to have their homes and lives disrupted by decontamination work, were skeptical about the remedial action program and refused to participate. The voluntary nature of the program also meant that program managers were unable to determine the full extent of the remedial work required. Project officials in 1978 knew that at least seven hundred buildings needed significant remedial action, but another six hundred locations suspected of being contaminated could not be fully evaluated because the owners or occupants refused to allow access to their properties or could not be contacted. This raised the possibility that contaminated properties would be sold to unsuspecting buyers after the termination of the project, perpetuating the indoor radiation pollution problem.[4]

The pre-UMTRCA Grand Junction remedial project officially ended in 1987, by which time the majority of the remedial action work was being conducted under the umbrella of the UMTRA Project. During the Grand Junction project's lifetime, contractors removed 140,000 tons of mill tailings and contaminated materials from under and around about six hundred structures. The problem of tailings contamination of other parts of the city, such as under sidewalks and around sewer lines, remained, requiring further remedial action under UMTRCA.[5]

Although the UMTRCA remedial action program was conceived on a scale far grander than the cleanup of Grand Junction, it was based on simple environmental engineering concepts. The tailings first needed to be dried, in some cases moved, and then covered to prevent erosion from wind, precipitation, and flood. The containment bunkers needed to be lined to prevent percolation of contaminants into the groundwater. A permanent soil cover had to be sufficiently thick to prevent radon gas emissions at rates over 20 picocuries per square meter per second. In 1980 the Nuclear Regulatory Commission, successor to the AEC, had required the protective cover be of sufficient durability to last "thousands of years," but the Reagan-era EPA reduced that requirement to "at least two hundred years." Rock or vegetative coverings would strengthen the earthen barrier, while the whole thing needed to be contoured to prevent runoff erosion. Generally, the regulations favored the removal of tailings to isolated locations for disposal when practical and economical, especially those tailings located in or near urban areas. In marked contrast to the earlier AEC approach to tailings, the regulations embodied the attitude that tailings must be interred with an eye to long-term health safety, regardless of the short-term inconveniences it might cause. They also reflected congressional expectations that the sites should be permanently reclaimed so that there would be no need for active, ongoing maintenance.[6]

Like the Grand Junction legislation, UMTRCA was enacted with high hopes of success but suffered serious implementation problems during its lifetime. Preliminary activities, such as surveying the sites, developing engineering solutions, soliciting public comment on the proposals, and acquiring disposal locations, took longer to accomplish than legislators originally anticipated. A greater obstacle in the path of effective implementation of UMTRCA, however, was the structure of the law itself. "The legislation was so poorly drafted," commented Larry Boggs, senior counsel for the American Mining Congress, "as to be unbelievable. The Congress was asleep [when it passed UMTRCA]."[7] Perhaps because it was pushed through during the closing hours of a Congress and was a compromise among several bills, the final version of the law was an awkward hybrid indeed. It required the cooperation of several federal agencies in setting the compliance standards and implementing the provisions of the act, resulting in jurisdictional confusion among the EPA, the DOE, the NRC, and state agencies. In addition, the federal agencies were slow to comply with the law, and there were few sanctions outside the legal system to compel agencies to act in a timely manner.

The greatest problem facing UMTRCA was its unfortunate timing. The first two years after passage of UMTRCA were largely devoted to assessment work and to creating the bureaucratic and technical infrastructure necessary to make the UMTRA Project a success. By the time actual earthmoving work was to begin, however, there had been a radical change in environmental philosophy in Washington. The election of Ronald Reagan to the presidency in 1980 represented a serious challenge to the environmental innovations of the 1970s.

Reagan and his environmental know-nothing supporters saw the regulatory systems that had been developed to oversee environmental regulations as a threat to corporate America. Pro-business, pro-development forces who financed the Republican victory feared "Big Government" and targeted environmentalists, and their causes, with special fury. To "set business free" and "get government off the backs" of business, Reagan and his advisers worked to turn back the clock on the nation's environmental protection agenda. Reagan used the Office of Budget and Management to shift federal finances away from regulatory duties. Budget cuts gutted environmental agencies; EPA staff was slashed 25 percent and its budget nearly 30 percent during Reagan's first two years in office. The new president also simply ignored enforcement of environmental regulations. Industrial trade associations were frequently invited to draft new environmental regulations more favorable to themselves. Reagan appointees, such as EPA Director Ann Gorsuch and Secretary of the Interior James Watt, were awarded their jobs because of their long-standing opposition to environmental initiatives and used their positions to undermine federal environmental protection initiatives. Environmental legislation and implementation ground to a halt. Under these conditions, the UMTRA Project faced an uphill battle during much of the 1980s.

UMTRCA's inopportune timing was evident in other ways as well. Uranium millers complained bitterly that UMTRCA was unjust because it placed on them the primary economic burden to clean up tailings at operating mills, at a cost industry estimated might reach a staggering $4.4 billion. Moreover, they noted, by 1978 about 80 percent of all tailings were located at active mills that were producing uranium or under NRC license. Their argument was persuasive. The legislative expectation underlying UMTRCA was for a bright economic future for uranium. The transition of the nation's atomic infrastructure from government control to private industry had been difficult during the 1960s. The growth in the atomic power industry slowed, especially during the late 1960s, as public opinion increasingly turned away from nuclear power. The situation changed dramatically during the early 1970s, however. Uranium prices rose during the decade, triggering a brief second uranium boom. UMTRCA legislators reasoned that the mill profits from the escalating uranium market would easily cover remediation costs.

Under the circumstances, their optimism can be understood. Several factors drove the surging yellowcake market during the 1970s, reinforcing optimistic predictions about the future of the uranium industry. OPEC's oil embargo in 1973 increased the cost of all American energy and spurred new interest in nuclear power. Skyrocketing oil prices made atomic energy appear to be more financially competitive with fossil fuel plants. The "energy crisis" renewed Cold War aspirations that America should be energy independent, free from the mercy of the OPEC oil states. Industry and government projections showed that American energy consumption would continue to rise and that atomic energy was the best alternative to meet the new demand. The price for uranium rose sharply as power companies scrambled to negotiate secure uranium delivery contracts for their new and existing reactors and hoarded their existing uranium inventories.[8] The American embargo on foreign uranium, first instituted in 1966, shut international uranium producers out of the world's best market. In response, they organized an international uranium cartel in 1972 to control production, distribution, and prices for uranium, virtually eliminating competition for yellowcake. By 1974 the cartel had levered the price of yellowcake to $40 per pound. Domestic yellowcake prices increased in relation to prevailing world prices.[9] Finally, the United States had a monopoly on uranium enrichment in the noncommunist world until 1977. In 1973 the government moved away from its traditional short-term enrichment agreements and instituted a new policy, insisting on long-term contracts to ease planning for enrichment services and maximize the benefit of its monopoly. Utilities, which were frantically planning new reactors to meet anticipated demand for nuclear power, had no choice but to accept these new conditions to avoid possible shortages in their supply of uranium. Consequently, the government's new enrichment policy forced power companies to accept extended contracts, committing them to long-term uranium purchases for existing reactors as well as those that were only on the drawing boards, thereby increasing demand for uranium well into the future and putting upward pressure on uranium prices.[10]

Fantastical predictions for future uranium output eclipsed the most productive years of the 1950s and 1960s. In 1976, for example, the Federal Energy Administration calculated that New Mexico's uranium yield alone would reach 65,000 tons annually—six times more than 1960, the most productive year for New Mexico's uranium industry. Tailings from that new production would reach 949 million tons, ten times more than what existed in New Mexico in 1977. The price for yellowcake was $6.41 per pound in 1973 but shot up to nearly $50.00 per pound in 1977 before settling down to about $43.00 per pound in 1978. In 1978 there were twenty-one mills operating in six western states, and some of the mills announced plans to expand to meet the projected uranium demand. Eleven new mills were expected to be operating by the late 1970s or early 1980s, and several more mills were in the planning stages. Forecast after forecast painted a bright future for nuclear power. Such economic growth, federal legislators believed, could easily absorb the cost of UMTRCA tailings management, not only the existing 130 million tons accumulated by the end of the 1970s, but the additional millions of tons of projected waste as well.[11]

Then the effervescent uranium market crashed. The uranium boom of the 1970s, founded more on speculation and wishful thinking than on practical economic reality, was short-lived. Although UMTRCA was passed in the midst of the boom, by the time the law became effective in any practical sense, the nuclear power industry was once again dying, the victim of Three Mile Island, public hostility, blistering criticism from environmentalists, lower oil costs, and the economic reality that nuclear power was more expensive than conventional sources. The American embargo on foreign uranium had been eased beginning in 1977 to meet the overly optimistic projected demand for uranium, further depressing prices as lower-cost foreign uranium flowed into the domestic market. Uranium companies had made deep financial commitments, betting on a future that never materialized. Few of the planned new mills ever opened, and those that did open after the passage of UMTRCA closed by the early 1980s. Even at the operating mills, as of 1985, employment had fallen 90 percent from its peak in the 1960s, from about 22,000 to 2,200. Uranium prices plunged from their late 1970s peak to $17 per pound by 1982 and below $10 per pound in 1989. Most domestic uranium ore was so low-grade that American producers simply could not compete at the depressed prices. Only five of the "active" uranium mills were in fact operating at all, usually at less than peak capacity. The remaining mills were in an industrial limbo; too unprofitable to operate but too expensive to close because of the UMTRCA cleanup requirements and too risky to close in the event that uranium prices rebounded. American mills had once provided half of all uranium produced worldwide, but by the mid-1980s they produced less than 25 percent of even domestic requirements. Consequently, the justification for placing Title II UMTRCA financial responsibility on the industry never materialized. Milling executives pointed out that the battered and barely surviving milling industry was in no position to shoulder UMTRCA's crushing remedial costs.[12]

Industry's complaints did not fall on deaf congressional or administrative ears, and political interference further hampered the already problematic law. Sympathetic lawmakers like Senators Alan K. Simpson (R-Wyoming) and Pete V. Domenici (R-New Mexico) and Representative Bill Richardson (D-New Mexico), emboldened by the antienvironmental and government reduction climate of the Reagan administration and the Republican-dominated Congress that came to power in 1981, offered several amendments to UMTRCA. They wanted to loosen the standards, extend the deadline for agencies to promulgate UMTRCA standards and regulations, and give mill operators greater flexibility in implementing the provisions of the law. In 1982 Congress amended an appropriations bill to prohibit the NRC from any spending to implement tailings regulations. For many politicians, especially those from western states contaminated by tailings, such action went too far. Although subsequent legislation restored the NRC's power in January 1983, the underlying hostility to UMTRCA was unmistakable.[13] Administration officials, representatives, and senators pressured agencies, especially the EPA, to delay implementation of the cleanup project. Millers generated sympathy for their cause by reminding legislators that the delays might postpone final mill closures and bolster the flagging economies of declining uranium mill towns in their states and districts. They also suggested that heavy cleanup costs might inhibit any future revival of the uranium industry in those communities.

Milling industry critics, on the other hand, maintained that cleanup was not prohibitively expensive to either the mills or their multinational parent companies. Electric utilities claimed that cleanup costs were factored into the price they paid for nuclear fuel. In any case, they alleged, the collapse of the domestic uranium market was in large measure the result of unreasonable business practices of the uranium industry itself that forced utilities to accept fixed amounts of uranium at high prices despite plant construction delays and cancellations. Those business practices acted to artificially support uranium prices at high levels, providing mill companies with the cash necessary to deal with their tailings. It was not their problem, power executives argued, if those reserves were now gone. Other critics noted that the uranium industry could easily have afforded comprehensive tailings management all along but chose to maximize their profits at the expense of the environment and public health. In 1959, for example, the cost of adequately disposing of tailings at the mill in Monticello, Utah, would have been about $0.64 per ton. At the same time, the uranium in that ton of ore, assuming the ore contained 1 percent uranium, was worth about $180.00. Public health officials feared that the new amendments might result in inadequate safety standards.[14]

The milling industry did not rely on political pressure alone to relieve its burden. It also turned to the federal courts, which increasingly provided the forum for the practical and ideological battles over mill tailings management. During the late 1970s, the NRC had designed a generic environmental impact statement outlining minimum recommended standards for tailings remediation. By 1980, three years before the EPA finally completed its standards as required by UMTRCA, NRC was in a position to issue rules implementing UMTRCA. Several mill companies challenged the NRC regulations in court, alleging that they were too strict. The companies also argued that the NRC had usurped EPA authority by issuing regulations to implement UMTRCA before the EPA's general standards were complete. The challenge was unsuccessful, but the loss did not deter further legal action on the matter.[15]

The Reagan-era EPA especially came under fire from environmentalists for its lax handling of its tailings responsibilities. In October 1983, nearly three years after the original deadline set by Congress, the EPA finally issued its mandatory standards for control and cleanup of the tailings sites. Few were happy with the standards. Industry believed they were too stringent. Environmentalists were outraged that the standards allowed ten times more radon emissions from reclaimed tailings than the standards previously issued by the NRC; they also criticized the EPA for failing to issue any emission standards for operating, unreclaimed tailings piles. The environmental groups and the milling industry challenged the EPA regulations. The Tenth Circuit Court of Appeals sided with environmentalists, chastised the EPA for its obstructionist behavior, and upheld the more stringent standards.[16]

During the late 1980s, mill companies that had produced uranium and thorium under the AEC procurement program sued the United States to recover their UMTRCA compliance costs. The plaintiffs sought recovery for costs of managing the tailings associated with the AEC purchases but not those arising from commercial sales to private industry. These so-called co-mingled tailings were common and posed difficult legal and administrative problems. Millers claimed that the government could not retroactively burden them with the costs of Title II remedial action because their original contracts did not include provisions for managing tailings on the scale mandated by UMTRCA. In a sweeping decision, the U.S. Court of Appeals affirmed a federal claims court dismissal of the plaintiffs' plea for equitable reformation of their uranium sales contracts. The court rejected all the plaintiffs' arguments, reaffirming the indisputable legislative intention to place the primary Title II UMTRCA burden on the companies. It declared that the NRC had discretion to shape unique policies to manage individual tailings sites and that the act granted the NRC the power to develop regulations that imposed economic hardship on millers. More significantly, the court's ruling, which turned back the strongest challenge to UMTRCA, upheld Congress's intent that industry solve the tailings problem once and for all at operating facilities.[17]

The legal validation of UMTRCA cleared the way for concrete measures to contain the tailings, but it soon became clear that the program faced serious financial problems. As the UMTRA Project slowly unfolded, the price tag for remedial action at the Title I abandoned tailings sites ballooned from the original estimate of $200 million. By the mid-1980s, just as the actual cleanup work was beginning, federal budget managers feared the costs might top $900 million. Their concerns were justified; the final cost for the surface phase of the UMTRA Project alone was ultimately about $1.5 billion, of which the federal government provided about $1.4 billion and states contributed $100 million. The DOE projected in 1995 that groundwater cleanup and monitoring, which will last until at least 2010, would cost an additional $497 million, of which $2.5 million would come from the states. In contrast, the total value of the uranium processed at all the uranium mills between 1948 and 1971 was about $3 billion.[18] The costs at privately owned, Title II sites also escalated. The surety bonds, intended to guarantee funds for site stabilization if a mill owner defaulted, totaled only about $300 million by 1985, far short of the actual costs of remedial work required. That administrators permitted a mill to defer cleanup until termination of its license meant that cleanup costs continued to mount even after the effective date for UMTRCA but before any mill company began its remedial work. Several companies faced a difficult dilemma: keep an unprofitable mill running to avoid paying remediation costs, or face bankrupting cleanup liability when the mill closed. Most companies opted to defer cleanup for as long as possible.[19]

In spite of these political, legal, and economic obstacles, remedial action slowly progressed. The DOE opened its UMTRA Project operations office in Albuquerque, New Mexico, during 1979 and hired contractors to carry out engineering research, remedial action, and environmental and health safety monitoring. A separate Grand Junction Projects office oversaw the vicinity property cleanup in that city. From 1980 until 1983 DOE spent $12.4 million on a comprehensive technology development program to address the specific engineering details of remedial action and to anticipate previously unforeseen facets of the tailings program.

When the EPA finally issued its environmental and health safety standards for the UMTRA Project in January 1983, it cleared the way for the DOE to begin surface remedial action.[20] It also officially started the clock on the seven-year deadline originally set for the project's completion. A few months later, the agency issued its standards for groundwater contamination. Groundbreaking on the first cleanup site was October 7, 1983, at Canonsburg, Pennsylvania. During that year, DOE contractors began work on vicinity properties in Canonsburg, Salt Lake City, and Grand Junction. In 1984 they began remediation at Shiprock, New Mexico, and Tuba City, Arizona. The following year, work began in Salt Lake City, and the Canonsburg project was completed in December. During 1986, cleanup began at Lakeview, Oregon, and Durango, Colorado, and the Shiprock program was completed in the fall. Mexican Hat, Utah, and Ambrosia Lake, New Mexico, were under way beginning in 1987.

When it became evident that the initial seven-year timetable could not be met, Congress amended UMTRCA in 1988 to extend the DOE's cleanup authority four years beyond the original deadline, until 1994. The amendment also mandated groundwater compliance, though it did not specify time restraints. Work began in Riverton, Wyoming; the two Rifle, Colorado, sites; Green River, Utah; and the Climax pile in Grand Junction. The next year, contractors initiated programs at Spook, Wyoming, and Monument Valley, Arizona, and completed the surface work at Salt Lake City, Spook, Lakeview, and Green River. Tuba City and Riverton were completed in 1990.

During the 1980s, aquifer restoration had not been a priority for the Reagan-era EPA and NRC, and mills were allowed to work on groundwater problems largely at their own pace. Most were happy to be relieved of the responsibility of timely and expensive aquifer restoration measures. Groundwater restoration standards varied considerably because the geology and hydrology of tailings sites differed, and most were overly optimistic. That leisurely pace ended in 1991, when the groundwater restoration phase of the UMTRA Project commenced in earnest with an estimated price tag of $1 billion. The DOE approached the problem of groundwater as a separate phase of the UMTRA Project in order to address the anticipated decades-long life of the program. In fact, several groundwater programs will not be complete until well into the next century.[21]

The 1990s signaled an end to the hesitant enforcement of UMTRCA. In 1991 the NRC and the EPA abandoned the open-ended approach to remediation and agreed to enforceable deadlines for abandoned sites. The EPA's goal was to have permanent covers in place on all abandoned tailings sites by the end of 1997. Under the agreement, mill companies faced specific timetables for reclamation of their tailings for the first time.[22] The new arrangements increased the pace of remedial action. Surface remediation at Lowman, Idaho, and the completion of work at Durango also commenced in 1991. In 1992 surface cleanup activity began at two more sites, Falls City, Texas, and Gunnison, Colorado. Lowman was completed in the spring.

Congress once again amended UMTRCA in 1992, extending the life of the UMTRA Project another two years, to 1996. In 1992 the disposal cell at Spook became the first UMTRA Project disposal site licensed by the NRC, the last step in the surface cleanup process. Under the terms of the NRC license, the Commission brought the site under its long-term surveillance plan. Ultimately, all the surface sites would come under NRC license control by 1998. The project at Naturita began in 1994, and remediation was completed at Monument Valley, Falls City, and Grand Junction. In 1995 work began at three Colorado locations, Maybell and the two Slick Rock sites, and work was completed at Mexican Hat, Ambrosia Lake, and Gunnison. On October 9, 1996, President Bill Clinton signed a bill that extended UMTRCA for the last time, until September 30, 1998. During 1996, the two Rifle sites were completed, as were the two Slick Rock locations in 1997. In 1998 the final projects, at Maybell and Naturita were completed, and by the end of September, all twenty-two sites were completed, and sixteen of the eighteen disposal cells were licensed by the NRC.[23] The UMTRA Ground Water Project is continuing into the twenty-first century.

It is appropriate that the DOE celebrated completion of the surface phase of the UMTRA Project in Grand Junction, the community that had played such a central role in the tailings controversy and which was one of the most contaminated communities of the atomic age. Yet Grand Junction was not unique. Towns like Green River and Mexican Hat, Gunnison and Rifle, Riverton, Lowman, and Falls River were unlikely settings for one of the nation's most controversial atomic pollution debates or the ambitious environmental remediation projects designed to remedy the contamination. Most had passed the first half of the century as small, close-knit communities, generally conservative, and fiercely proud of their western heritage. They welcomed the uranium mills during the Cold War for the jobs they brought but equally because the mills symbolized the nation's commitment to protecting freedom and democracy in a hostile world. For years the townspeople tolerated the inconveniences, and dangers, of the mills in the name of prosperity and patriotism, but when the mills died, the communities continued to be plagued by the troublesome legacy of the nation's quest for uranium independence. For more than two decades, Grand Junction endured the laborious and inconvenient process of having soil under houses, public and private buildings, and sidewalks and streets removed, hauled away, and replaced with uncontaminated earth. For years Durango and Grand Junction residents heard the steady rumble of trucks that carried the tailings to the disposal cells. In Salt Lake City, trains carried the Vitro tailings to a disposal cell west of the Great Salt Lake. These sights and sounds, repeated at one time or another in all the contaminated mill towns, served as reminders of the price these communities paid to secure the nation's uranium.

Yet for all its success, perhaps the most salient feature of the history of UMTRCA was the lassitude with which it was carried out. For all of its good intentions, it was a law out of step with the times. The optimism about tailings remediation that prevailed in Congress when UMTRCA became law evaporated only a few years later. The energy crisis of the early 1970s offered hope of reviving the sluggish uranium industry, but the near-disaster at Three Mile Island nuclear power plant and the powerful environmental spirit of the times together snuffed out the future for atomic power in America. After the uranium market crash in 1980, federal and state regulators discovered that their responsibility had fundamentally changed. Rather than direct a thriving industry to reclaim its tailings legacy, their main responsibility was to oversee the death throes of an industry. That bleak economic reality was not what the UMTRCA policy makers had envisioned. Regulators struggled to adjust to their new responsibility, and the implementation of UMTRCA suffered as a result. Mill companies saw the UMTRCA requirements as the final nail in their industrial coffin and were reluctant to fulfill their obligations.

Administrators and politicians worked to reformulate UMTRCA to meet the changed economic conditions. The newly elected Republican president and Senate made those changes easier and, in light of the antigovernment, pro-business climate that prevailed in Washington during the 1980s, those adjustments benefited the uranium industry. Like much of the Reagan-era approach to environmental issues, the implementation of UMTRCA was seemingly contradictory. On the one hand, administrators adopted a casual, even obstructionist attitude about when reclamation would begin, the pace at which it would proceed, and the standards required to be met during reclamation. Long-term safety concerns were subordinated to short-term industrial expectations. On the other hand, legislators and the administration used the power of the government to directly support the uranium industry. They used their influence to reshape UMTRCA provisions to ease the burden on private industry. Finally, although millers had lost their court battle to recover costs for reclaiming tailings generated under federal contract, in 1992

Congress agreed to reimburse millers for disposing of 55 million tons of those tailings.[24] The action effectively rewrote the original premise of UMTRCA, shifted the environmental burden onto taxpayers, and proved once again that uranium production in America was economically unjustified without substantial government assistance. Given the level of political and administrative hostility to UMTRCA during its formative years, it is perhaps surprising that it ever survived at all.

Over half a century has passed since the United States made its Faustian bargain with the atom. The apocalyptic examples of Hiroshima and Nagasaki and the harrowing experience with the Cold War compelled atomic scientists to imagine peaceful uses for the new knowledge that would outweigh its destructive features. Policy makers and technocrats assumed that atomic fission would significantly supplement fossil fuels, perhaps eventually replace them, as the nation's primary energy source. They championed nuclear power on the ideological grounds that the new technology would lead to a utopian future. Notably absent from their plan was a comprehensive and environmentally sound program to manage front-end nuclear waste that would inevitably follow widespread use of fission power.

There are many contradictions in the nation's atomic experiment; the history of uranium mill tailings provides only one example. The tailings story juxtaposes two antithetical views about the national welfare. On the one hand, atomic technocrats and policy makers were determined to make atomic energy an integral part of the nation's electrical generating infrastructure. Although the AEC issued exposure radiation standards, the agency's promotional zeal meant that it was predisposed to downplay potential health hazards that might interfere with the success of that mission. On the other hand, public health and environmental experts were much more willing to view radioactive waste from uranium tailings as a potential danger and to err on the side of caution, even if it interfered with the nation's nuclear dreams.

The 1954 Atomic Energy Act, high tide of the federal policy to develop domestic atomic power, is the symbol of the peaceful atom. After the mid-1950s the government officially acknowledged that nuclear power was a national policy goal. The federal government began building a partnership with private industry to develop commercial applications of atomic energy. At the same time, the act reaffirmed federal responsibility for protecting the public from the hazards associated with the anticipated growth of the atomic power industry.

During the 1960s, environmental quality eclipsed atomic energy as one of the most urgent public policy issues in the United States. In contrast to the development of nuclear power, a policy originated at the highest levels of the federal government and within the boardrooms of corporations, interest in environmental matters arose from the citizenry, who exerted relentless pressure on policy makers to consider the environmental consequences of their decisions. The roots of this new environmental consciousness lay, to great degree, in the fears about atomic energy that underscored American culture since the end of World War II, especially anxiety about nuclear fallout, reactor safety, and the front end of the uranium fuel cycle. The most important feature of this new environmental movement was the idea that "pollution" threatened people's health, homes, and communities. The homocentric emphasis on pollution was an awakening for a broad stratum of the American public that modern technologies, like atomic energy, were hardly as benign as advertised but rather posed significant health and safety dangers. The apparent insensitivity of industry and government's tepid response to the public's atomic pollution fears during the early 1960s led many Americans to conclude that private industry and the government especially were environmental villains.

At the same time the environmental movement commanded increasing public attention, the national hunger for electricity compounded questions about environmental integrity. The urgency for alternative fuel sources became even more acute after the energy crisis of the early 1970s. The dual demands for abundant energy and a clean environment put policy makers in a dilemma. Fossil fuel facilities could meet the nation's power demands but were significant polluters, annually spewing millions of tons of hazardous chemicals into the air. Ironically, the new environmental concerns about pollution offered the atomic proponents a seemingly easy solution. Reactors, they claimed, could provide the energy the nation demanded with less air pollution than fossil fuel facilities produced. Federal policy makers, particularly within the AEC, remained enthusiastic about the possibility that atomic energy could solve the ecology and energy conundrum. Increasingly, however, that official commitment to fission power was at odds with public environmental expectations. Events like the uranium mill tailings crisis eroded public confidence in nuclear technology and the government bureaucrats responsible for controlling the atomic genie. Ultimately, the power of public opinion overwhelmed the best efforts of government and industry, and atomic energy died as a viable alternative to fossil fuels.

With hindsight, we see that uranium mill tailings were neither as threatening as some critics imagined nor as benign as the technocrats and nuclear supporters assumed. It is difficult to assess the actual consequences of tailings pollution. The UMTRA Project cost taxpayers millions of dollars. Uranium companies spent millions more to stabilize piles that did not qualify under UMTRCA. In 1992 the policy analyst W. Kip Viscusi estimated the benefits and costs of tailings control: regulating mill tailings saves about 2.1 lives per year at $53 million per life.[25] These figures demonstrate that while the risk from tailings was perhaps low and remediation costs great, the hazard was nevertheless real. Yet, as nearly four decades of debate over uranium mill tailings pollution shows, the lasting impact of the crisis cannot be measured simply in money or theoretical human lives. The tailings issue was, instead, a parable about the power of public perspectives and beliefs to shape environmental policy. The story of uranium tailings policy and environmental pollution illustrates the bridge that spans the gulf between the early years of the atomic era, when everything seemed possible and no scientific or technical problem appeared too difficult to solve, to the time when environmental idealism bred suspicion of the atom. As part of the story of the post–World War II environmental movement, it exemplifies some of the shortcomings of the nation's nuclear policy. First, the tailings crisis sowed seeds of public doubt about the safety of atomic energy. Americans learned that nuclear power, the alleged logical successor and "clean alternative" to fossil fuel plants, created significant amounts of hazardous waste. The public believed the atom's critics, and reassurances by federal and industrial experts, and legalistic denials of responsibility, could not overcome growing public suspicions. Second, the hesitant federal response to the tailings pollution problem, however justified on legal and regulatory grounds, undermined public faith in the AEC's commitment to curb atomic pollution and protect public health. This loss of confidence in the federal atomic agency's ability to oversee what growing numbers of Americas saw as a dangerous technology contributed inevitably to a declining national commitment to atomic energy as an acceptable alternative to fossil fuels and hastened the demise of the atomic energy industry.

Third, the tailings crisis exposed the inherent weakness of the AEC's dual mandates of atomic promotion and safety and, indirectly, raised doubts about the nation's legislative formula for controlling the atom. The AEC, charged with shielding the public from atomic dangers, increasingly viewed environmental oversight as an obligation secondary to its atomic promotion and national security responsibilities. While it paid lip service to environmental and health safety matters, AEC officials regularly downplayed evidence pointing to possible health dangers that might erode national faith in nuclear power. They insisted that their statutory mandate to ensure atomic safety did not extend to hazards materials, like tailings, that were most commonly associated with the front end of the nuclear fuel cycle. Finally, UMTRCA, like much of the environmental legislation of the 1960s and 1970s, was a remedial action law, designed to address a serious environmental problem after the damage had been done. While it was commendable that Congress allocated funds for the cleanup program, it is a central tragedy of the tailings story that the cost to taxpayers and millers to dispose of the tailings was far greater than it would have been to effectively manage the tailings in the first place, especially once the potential health threat from the tailings became clear during the late 1950s.

Politics, according to the cliché, is the art of the possible, and the same is true for environmental policy making. Although states succeeded in forcing the federal government to pay the lion's share of the cleanup costs, it was never likely that UMTRCA would please all the different parties who had a stake in atomic power, especially after the difficult economic and political climate of the 1980s. Despite its structural shortcomings, ideological disputes among environmentalists, industry, community representatives, and the DOE, its technical implementation mistakes, and its time and cost overruns, UMTRCA was a milestone in the history of America's atomic program and the development of the environmental movement. It was one of the few significant amendments to the Atomic Energy Act and thus signaled a fundamental shift in the nation's official atomic energy policy goals. UMTRCA, although perhaps thirty years too late, was a clear congressional rejection of the business-as-usual approach to atomic energy and marked a new era in which the federal government accepted responsibility, however grudgingly, for solving a serious environmental pollution problem that it was instrumental in creating. It was a definitive admission by Congress that nuclear power entailed risks that the public was unwilling to take.

UMTRCA and the completion of the surface phase of the UMTRA Project ended the tailings pollution problem that had so dramatically intruded into the lives of mill community residents. The cleanup was not easy, and like Faust himself Americans paid a great price for their acquired wisdom. Perhaps, however, UMTRCA enabled the United States to exorcise at least some of the ghosts of its atomic past.

Notes

Introduction

1. U.S. Department of Energy, Office of Environmental Management, UMTRA: *Fiscal Year 1995 Annual Report to Stakeholders* (Washington, D.C.: USDOE, 1995); Holger Albrethsen, Jr., and Frank E. McGinley, *Summary History of Domestic Uranium Procurement under U.S. Atomic Energy Commission Contracts* (Grand Junction, Colo.: USDOE Area Field Office, September 1982).
2. Stephen Hilgartner, Richard C. Bell, and Rory O'Connor, *Nukespeak: Nuclear Language, Visions, and Mindset* (San Francisco: Sierra Club Books, 1982), 190.
3. James Miller, *Democracy in the Streets* (New York: Simon and Schuster, 1987), 330, 364–65.
4. Daniel Ford, *The Cult of the Atom: The Secret Papers of the Atomic Energy Commission* (New York: Simon and Schuster, 1982), 50; Steven Mark Cohn, *Too Cheap to Meter: An Economic and Philosophical Analysis of the Nuclear Dream* (Albany: State University of New York Press, 1997), 107.
5. Paul Boyer, *By the Bomb's Early Light: American Thought and Culture at the Dawn of the Atomic Age* (New York: Pantheon Books, 1985), 15.
6. Throughout this book, I use the term "technocrat" to describe a government official or scientist who, because of his or her expertise in a highly technological field, has the ability to understand and exert control over aspects of that technology and related public policy in a manner unavailable to government officials lacking the expertise and who advocates solutions to social issues based on that technology. "The technocrat," notes Peter Taylor, "believes that he can handle social complexity in a value-free manner, maintaining a distance from specific interests and political details, and that through such nondependency and disengagement he can best serve all. But it is typical of social philosophies framed in terms of universal interests that their proponents hold a special place in the proposed social organization." Peter J. Taylor, "Technocratic Optimism, H. T. Odum, and the Partial Transformation of Ecological Metaphor after World War II," *Journal of the History of Biology* 21 (1988): 215.
7. See generally, Brian Balogh, *Chain Reaction: Expert Debate and Public Participation in*

American Commercial Nuclear Power, 1945–1975 (Cambridge: Cambridge University Press, 1991).
8. Cohn, *Too Cheap to Meter*, 88.
9. In 1970 Dr. Philip Handler, president of the National Academy of Sciences, noted, "It is essentially impossible to find more than a small handful of such experts who have not, in relatively recent times, had significant support—either research grants or actual employment—from the U.S. Atomic Energy Commission." H. Peter Metzger, *The Atomic Establishment* (New York: Simon and Schuster, 1972), 26. Nine years later, Constance Holden criticized the close relationship between the AEC and the scientific establishment: "The radiation research community has lived almost entirely off the energy and defense establishments. The situation is conducive to a monolithic approach the research.... It also means that for anyone seeking objective scientific advice it is practically impossible to find someone knowledgeable who was not trained with AEC money." Constance Holden, "Low-Level Radiation: A High-Level Concern," *Science* 204 (April 13, 1979): 156.
10. Cohn, *Too Cheap to Meter*, 90.
11. Ibid., 88–91. "In defense of the AEC, it should be noted that given the Commission's optimistic technical assumptions, the AEC's regulatory decisions were sometimes conservative." Ibid., 91. See generally, Louis Gwin, *Speak No Evil: The Promotional Heritage of Nuclear Risk Communication* (New York: Praeger, 1990).
12. The Department of Labor led the way in creating radiation safety standards for uranium mines in 1967. A few years later, the National Environmental Policy Act transferred many radiation safety and regulatory functions to the Environmental Protection Agency. Cohn, *Too Cheap to Meter*, 92.
13. Walter C. Patterson, "Chernobyl—The Official Story," *Bulletin of the Atomic Scientists* 42 (November 1986): 36.
14. Balogh, *Chain Reaction*, 16.
15. Ibid., 1–20, 171–222.
16. Robert C. Merritt, *The Extractive Metallurgy of Uranium* (Golden: Colorado School of Mines Research Institute, 1971), prepared under contract with the U.S. Atomic Energy Commission.
17. David Riccitiello et al., "Uranium Mining and Milling," *Workbook* 4 (1979): 222; *Environmental Standards for Uranium and Thorium Mill Tailings at Licensed Commercial Processing Sites*, 48 Fed. Reg. 45, 45,926–45,928 (1983); Luther J. Carter, *Nuclear Imperatives and the Public Trust* (Washington, D.C.: Resources for the Future, 1987), 12; "Landscaping (Industrial Strength)," *Nuclear Energy* (Second Quarter, 1993): 9–11.

It is difficult to determine precisely the amount of tailings, and authors have offered various estimates. The likeliest conservative estimate is about 140 million tons. See Metzger, *The Atomic Establishment*, 162; Rodger Rapoport, "The Trouble with 90.5 Million Tons of Radioactive Tailings," *Los Angeles Times (West Magazine)*, April 12, 1970, 10; "Uranium Mill Tailings: Congress Addresses a Long-Neglected Problem," *Science* 202 (October 13, 1978): 191; Corinne Brown and Robert Monroe, *Time Bomb: Understanding the Threat of Nuclear Power* (New York: William Morrow, 1981), 81. Worldwide, there may be as much as six billion tons of tailings. Rosalie Bertell, *No Immediate Danger: Prognosis for a Radioactive Earth* (London: Women's Press Limited, 1985), 81.
18. "Half-life" refers to the time it takes for the number of radioactive nuclei originally present in a sample to decrease through radioactive decay to one-half their original

19. J. J. Swift, J. M. Hardin, and H. W. Calley, *Potential Radiological Impact of Airborne Releases and Direct Gamma Radiation to Individuals Living Near Inactive Uranium Mill Tailings Piles*, PB258-166, U.S. National Technical Information Service (1976); Glen A. Watford and John A. Wethington, Jr., "Radiological Hazards of Uranium Mill Tailings Piles," *Nuclear Technology* 53, no. 3 (June 1981): 295–301. A curie is a measure of radioactivity and refers to the amount of any nuclide that undergoes $3.7(10^{10})$ radioactive disintegrations per second.
20. Sigismund Peller, "Lung Cancer among Mine Workers in Joachimsthal," *Human Biology* 11 (1939): 130–43; W. C. Hueper, *Occupational and Environmental Cancer of the Respiratory System* (Springfield, Ill.: Charles C. Thomas, 1942), 435–56; Egon Lorenz, "Radioactivity and Lung Cancer: A Critical Review of Lung Cancer in the Miners of Schneeberg and Joachimsthal," *Journal of the National Cancer Institute* 5 (August 1944): 1–15; "Chronology of Uranium Mining Health Protection Activities," November 1970 (SECY-934, January 22, 1971), Box 7822, United States Atomic Energy Commission Secretariat File, United States Department of Energy Archives, Germantown, Md.
21. Metzger, *The Atomic Establishment*; Ralph Nader, *The Menace of Atomic Energy* (New York: Norton, 1979); Donald Bartlett and James Steele, *Forevermore: Nuclear Waste in America* (New York: Norton, 1985); Carter, *Nuclear Imperatives and the Public Trust*; J. Samuel Walker, *Containing the Atom: Nuclear Regulation in a Changing Environment, 1963–1971* (Berkeley: University of California Press, 1992).
22. Eric A. Nordlinger, *On the Autonomy of the Democratic State* (Cambridge, Mass.: Harvard University Press, 1981), 3, 15.
23. Alonzo P. Plough and Sheldon Krimsky, "The Emergence of Risk Communication Studies: Social and Political Context," *Science, Technology & Human Values* 12 (Summer 1987): 4.
24. David E. Lilienthal, *Change, Hope and the Bomb* (Princeton: Princeton University Press, 1963), 129.
25. Arjun Makhijani, Stephen I. Schwartz, and William J. Weida, "Nuclear Waste Management and Environmental Remediation," in Steven I. Schwartz, ed., *Atomic Audit: The Costs and Consequences of U.S. Nuclear Weapons since 1940* (Washington, D.C.: Brookings Institution Press, 1998), 380.
26. Ibid., 354. As of 1996 the Uranium Mill Tailings Remedial Action Program accounted for about 0.7 percent of the total cost for nuclear waste management and environmental remediation.

Chapter 1

1. Earle R. Caley, "The Earliest Known Use of a Material Containing Uranium," *Isis* 38, pts. 3, 4 (1947–48): 190–93. Caley cites "Analyses of Green and Blue Glass from the Posilipan Mosaic," *Archaeologia* 43 (1912): 99–108.
2. Herman Fleck, "A Series of Treatises on the Rare Metals," *Proceedings of the Colorado Scientific Society* (vol. 11) (Denver: Colorado Scientific Society, 1916), 154; L. O. Howard, "Development of Our Radium Bearing Ores," *Salt Lake Mining Review* 15 (February 28, 1914): 14; Herman Fleck and William G. Haldane, "A Study of the Uranium and Vanadium Belts of Southern Colorado," *Report of the Colorado State Bureau of Mines for the Years 1905–6* (Denver: Colorado State Bureau of Mines, 1907), 47–50; Thomas M. McKee, "Early Discovery of Uranium Ore in Colorado," *Colorado Magazine*

32 (July 1955): 192; Larry L. Meyer, "The Time of the Great Fever: U-Boom on the Colorado Plateau," *American Heritage* 32 (June–July 1981): 75; United States Vanadium Company, *Mesa Miracle in Colorado, Utah, New Mexico, Arizona* (New York: United States Vanadium Company, Union Carbide and Carbon Corporation, 1952), 10. Scientists soon understood that Klaproth had only isolated an oxide of uranium, and in 1841 Eugene Peligot obtained the metal itself. Joseph J. Katz and Eugene Rabinowitch, *The Chemistry of Uranium (Part 1): The Element, Its Binary and Related Compounds* (New York: McGraw-Hill, 1951), 122.

3. H. Carrington Bolton, "Index to the Literature of Uranium, 1789–1885," *Annual Report of the Board of Regents of the Smithsonian Institution, Part I (to July, 1885)* (Washington, D.C.: Government Printing Office, 1886), 915–46; Harold C. Hodge, "A History of Uranium Poisoning (1824–1942)," in H. C. Hodge, J. N. Stannard, and J. B Hursh, eds., *Uranium, Plutonium and Transplutonic Elements, Handbook of Experimental Pharmacology* (vol. 36) (Berlin: Springer-Verlag, 1973), 5–12. As late as 1860 investigators were conducting uranium experiments using human subjects, although there are no reports of fatalities. Uranium salts were so damaging to internal organs that they became the agent of choice for producing experimental acute nephritis in test animals.

4. Hodge, "History of Uranium Poisoning," 11. The lack of evidence about human injuries is probably due in part to the small homeopathic doses that became popular in higher dilution formulas.

5. Dan Kline and Ward Lloyd, eds., *The History of Glass* (London: Orbis, 1984), 174–75; Francis L. Pittman, "The Direct Production of Uranium Steel" (M.S. thesis, Colorado School of Mines, 1914), 22.

6. Uranium coloring powders could be used to produce black and six shades of yellow ceramic glazes. Richard B. Moore and Karl L. Kithil, "A Preliminary Report on Uranium, Radium, and Vanadium," *United States Bureau of Mines Bulletin*, no. 70 (Washington, D.C.: Government Printing Office, 1913), 58. See also Richard B. Moore, "Uranium and Vanadium," in G. A. Roush, ed., *The Mineral Industry: Its Statistics, Technology and Trade during 1920* (New York: McGraw-Hill, 1921), 708; R. D. George, *Common Minerals and Rocks: Their Occurrence and Uses* (*Colorado Geological Survey Bulletin*, no. 12) (Denver: Eames Bros., State Printers, 1917), 189; McKee, "Early Discovery of Uranium Ore in Colorado," 192; and Joseph Hyde Pratt, "Mineral Resources," *Twenty-first Annual Report of the United States Geological Survey to the Secretary of the Interior 1899–1900* (Washington, D.C.: Government Printing Office, 1901), 309; *Engineering and Mining Journal* 68 (1899): 310. In 1857 articles appeared in the *Liverpool Photographic Journal* and *Humphrey's Journal of the Daguerreotype* discussing the use of uranium for photography, and many similar articles appeared in following years in several European journals. Bolton, "Index to the Literature of Uranium," 922; Pittman, "Direct Production of Uranium Steel," 4.

7. Kathleen Bruyn, *Uranium Country* (Boulder: University of Colorado Press, 1955), 9–10.

8. Pratt, "Mineral Resources," 308; Wyatt Malcolm, *Notes on Radium-Bearing Minerals* (Canada Department of Mines, Prospector's Handbook, no. 1) (Ottawa: Government Printing Office, 1914), 4–25; Fleck, "A Series of Treatises on the Rare Metals," 157; H. W. Gillett and E. L. Mack, "Preparation of Ferro-uranium," *United States Bureau of Mines Technical Paper 177* (Washington, D.C.: Government Printing Office, 1917), 5–8; L. M. Dennis, "Uranium," in Richard P. Rothwell, ed., *The Mineral Industry: Its Statistics, Technology and Trade, 1897* (New York: Scientific Publishing Company,

1989), 654; J. Baxeres de Alzugaray, "Manufacture and Metallurgy of Ferro-vanadium," *Mining World* (June 24, 1905): 659–60; *Salt Lake Mining Review* 15 (February 28, 1914): 14; Pittman, "Direct Production of Uranium Steel," 22–24.

9. Fleck, "A Series of Treatises on the Rare Metals," 153–56; David T. Day (U.S. Department of the Interior, U.S. Geological Survey), *Mineral Resources of the United States, Calendar Year 1901* (Washington, D.C.: Government Printing Office, 1902), 268–70; Bruyn, *Uranium Country*, 31–32; Gordon Kimball, "Discovery of Carnotite," *Engineering and Mining Journal* 77 (June 16, 1904): 956.

10. Fleck and Haldane, "Study of the Uranium and Vanadium Belts of Southern Colorado," 48–50; Fleck, "A Series of Treatises on the Rare Metals," 156–58; "United States Mineral Production in 1897," *Engineering and Mining Journal* 65 (May 28, 1898): 638; U.S. Geological Survey, *Mineral Resources of the United States, Calendar Year 1901* (Washington, D.C.: Government Printing Office, 1902), 270; U.S. Geological Survey, *Mineral Resources of the United States, Calendar Year 1902* (Washington, D.C.: Government Printing Office, 1903), 15, 287; U.S. Geological Survey, *Mineral Resources of the United States, Calendar Year 1903* (Washington, D.C.: Government Printing Office, 1904), 309.

11. U.S. Geological Survey, *Mineral Resources of the United States, Calendar Year 1903*, 23, 309; *Calendar Year 1904*, 21, 343–46; U.S. Geological Survey, *Mineral Resources of the United States, Calendar Year 1905* (Washington, D.C.: Government Printing Office, 1906), 15, 413–14.

12. U.S. Geological Survey, *Mineral Resources of the United States, Calendar Year 1901*, 270; *Calendar Year 1902*, 287; *Calendar Year 1903*, 23, 308–9; *Calendar Year 1904*, 21; *Calendar Year 1905*, 413; U.S. Geological Survey, *Mineral Resources of the United States, Calendar Year 1906* (Washington, D.C.: Government Printing Office, 1907), 526, 539–40; U.S. Geological Survey, *Mineral Resources of the United States, Calendar Year 1907* (Washington, D.C.: Government Printing Office, 1908), 722; U.S. Geological Survey, *Mineral Resources of the United States, Calendar Year 1908* (Washington, D.C.: Government Printing Office, 1909), 10, 742, 748; U.S. Geological Survey, *Mineral Resources of the United States, Calendar Year 1909* (Washington, D.C.: Government Printing Office, 1910), 587; Fleck, "A Series of Treatises on the Rare Metals," 156–59; Bruyn, *Uranium Country*, 49–50.

13. "The Biological Effects of Radium," *Science* 33 (June 30, 1911): 1001–5; Carroll Chase, "American Literature on Radium Therapy Prior to 1906," *American Journal of Roentgenology and Radium Therapy* 8 (1921): 766–67; Eve Curie, *Madam Curie* (Garden City, N.Y.: Doubleday, Doran, 1938), 199. For a sample of literature, see Robert Abbe, "Subtle Power of Radium," *Transactions of the American Surgical Association* 22 (1904): 253–62; Louis Wickham and Paul Degrais, "Radium: Its Uses in Cancer and Other Diseases," *Contemporary* 98 (August 1910): 174–88; William A. Pursey, "Biological Effects of Radium," *Science* 33 (June 30, 1911): 1001–5; "Action of Radium upon the Embryo," *Scientific American*, Suppl. 72 (October 7, 1911): 235; James M. Davidson, "Vital Effects of Radium and Other Rays," *Nature* 88 (February 29, 1912): 600–602; "Recent Contributions to Radium Therapy, III," *Radium* 4 (October 1914): 74–80; Emile F. Krapf, "Recent Investigations on the Use of Radium for Malignant Diseases," *Radium* 1 (May 1913): 10–14.

14. H. E. Bishop, "The Present Situation in the Radium Industry," *Science* 57 (March 23, 1923): 341; Edward R. Landa, "Buried Treasure to Buried Waste: The Rise and Fall of the Radium Industry," *Colorado School of Mines Quarterly* 82 (Summer 1987): 1; Malcolm, *Notes on Radium-Bearing Minerals*, 25–26.

15. Fleck gives slightly different figures: as of 1914, ore averaging 2.5 percent uranium oxide cost about $42 per ton to mine, plus about $20 per ton to ship to Europe, leaving only about $18 per ton profit. Fleck, "A Series of Treatises on the Rare Metals," 174–75; U.S. Congress, House, Committee on Mines and Mining, *Radium Hearings on H.J. Res. 185 and 186*, 63d Cong., 2d sess., January 19–28, 1914, 166 (hereafter, *Radium Hearings*); Charles L. Parsons, "Our Radium Resources," *Science* 38 (October 31, 1931): 617.

16. Bruyn, *Uranium Country*, 42–43; Charles L. Parsons to Archibald Douglas, September 18, 1918, Box 373, Bureau of Mines, Record Group 70, General Classified File, General Correspondence File, National Archives and Records Administration, Washington, D.C. (hereafter Bureau of Mines Papers).

17. John S. MacArthur, "The Radium Industry and Reconstruction," *Engineering and Mining Journal* 107 (April 5, 1919): 605–6. Luminous paint originated as early as the seventeenth century. However, the ability of all such compounds to glow depended on previous exposure to a light source. The radioactivity of radium allowed for the creation of permanent and self-luminous compounds. A. T. Parsons, "Radium, with Special Reference to Luminous Paint," *Journal of the Oil and Colour Chemists' Association* 12 (January 1929): 3; Daniel Lang, "A Most Valuable Accident," *New Yorker* 35 (May 2, 1959): 49; James E. Lounsbury, "Famous Pittsburgh Industries: The Standard Chemical Company of Pittsburgh, Pa.," *Crucible* 22 (June 1938): 134; U.S. Bureau of Mines, *Annual Report to the Secretary of the Interior for the Fiscal Year 1918* (Washington, D.C.: Government Printing Office, 1919), 78; Wallace Savage, "Radioactive Luminous Materials," *Chemical and Metallurgical Engineering* 19 (September 28, 1918): 515–17; Charles H. Viol and Glenn D. Krammer, "The Application of Radium in Warfare," *Transactions of the American Electrochemical Society* 32 (1918): 381–90.

18. Lounsbury, "Famous Pittsburgh Industries," 134; Richard B. Moore, "Radium," in G. A. Roush, ed., *The Mineral Industry: Its Statistics, Technology and Trade during 1920* (New York: McGraw-Hill, 1921), 615–19; Charles H. Viol, "Radium Production," *Science* 49 (March 7, 1919): 227–28; MacArthur, "The Radium Industry and Reconstruction," 605–6; "Radium in the Home," *Literary Digest* 82 (September 13, 1924): 78–81.

19. *Radium Hearings*, 56.

20. Moore and Kithil, "A Preliminary Report," 58; Gillett and Mack, "Preparation of Ferro-uranium," 13; Robert M. Keeney, "The Manufacture of Ferro-alloys in the Electric Furnace," *Transactions of the American Institute of Mining Engineers* 140 (1918): 1365; Robert M. Keeney, "Uranium and Vanadium," in G. A. Roush, ed., *The Mineral Industry: Its Statistics, Technology and Trade during 1917* (New York: McGraw-Hill, 1918), 721–25.

21. U.S. Geological Survey, *Mineral Resources of the United States, Calendar Year 1916* (Washington, D.C.: Government Printing Office, 1919), 807; Arthur L. Miller, "Personal Reminiscences of the Early History of the Radium Extraction Industry in the U.S.A.," *Argonne National Laboratory Report ANL-7461* (July 1968): 98; Moore, "Radium," 709; Cyril G. Hopkins and Ward H. Sachs, "Radium Fertilizer in Field Tests," *Science* 41 (May 14, 1915): 732–35; R. R. Ramsay, "Radium Fertilizer," *Science* 42 (August 13, 1915): 219; MacArthur, "The Radium Industry and Reconstruction;" Louis D. Huntoon to Archibald Douglas, August 3, 1921, Box 728, Bureau of Mines Papers.

22. Bishop, "The Present Situation," 341–45; "$50 Milligram Price Spurs Search for Radium," *Business Week* (September 28, 1932): 20.

23. Bishop, "The Present Situation," 341–45; Camille Matignon, "The Manufacture of Radium," *Annual Report of the Board of Regents of the Smithsonian Institution, 1925* (Washington, D.C.: Government Printing Office, 1926), 233; Richard B. Moore, "Radium," in G. A. Roush, ed., *The Mineral Industry, Its Statistics, Technology, and Trade, 1922* (New York: McGraw-Hill, 1923), 617–19; Richard B. Moore, "Uranium and Vanadium," in G. A. Roush, ed., *The Mineral Industry, Its Statistics, Technology, and Trade, 1922* (New York: McGraw-Hill, 1923), 711–12; Frank L. Hess, "Radium, Uranium and Vanadium," in G. A. Roush, ed., *The Mineral Industry, Its Statistics, Technology, and Trade, 1926* (New York: McGraw-Hill, 1927), 601.
24. Edward R. Landa, "The First Nuclear Industry," *Scientific American* 247 (November 1982): 192; U.S. Bureau of Mines, *Bureau of Mines Bulletin,* no. 650 (Washington, D.C.: Government Printing Office, 1970), 672.
25. Herman Goodmen, "The Romance of Radium," *Medical Journal and Record* 81 (February 19, 1930) (New York: A. R. Elliot, 1930): 190–91; Curie, *Madam Curie,* 198. The first U.S. fatality from radiation was Clarence M. Dally, a glassblower and assistant to Thomas A. Edison during Edison's X-ray fluoroscopy experiments. E. R. N. Grigg, *The Trail of the Invisible Light* (Springfield, Ill: Charles C. Thomas, 1965), 14–15, 49, 783, 919; Bruyn, *Uranium Country,* 48; Jack Schubert and Ralph E. Lapp, *Radiation: What It Is and How It Affects You* (New York: Viking Press, 1957), 112–13; Lewis E. J. Roberts, *Nuclear Power and Public Responsibility* (Cambridge: Cambridge University Press, 1984), 30–33.
26. Robert A. Millikan, "The Significance of Radium," *Bulletin of the California Institute of Technology* 29 (June 1921): 3–21.
27. R. D. Evens, "Radium Poisoning: A Review of Present Knowledge," *Health Physics* (1980): 880–82; McKee, "Early Discovery of Uranium Ore in Colorado," 201. "Considerable amount" is a relative term; one dial painter who died of radium poisoning had only six-millionths of a gram in her body. As little as one half-millionths of a gram stored in the bones has proved lethal. Radium is eliminated naturally from the body but at a very low rate: it requires about forty-five years to eliminate one-half of the radium from the system. Many victims of radium poisoning did not live even that long. Schubert and Lapp, *Radiation,* 109–14. *New York World,* May 10, 1928; "Malignant Growths Resulting from Exposure to Radioactive Substances," *American Journal of Public Health and the Nation's Health* 22 (July 1932): 760–61; "Health Aspects of Radium Dial Painting," *American Journal of Public Health and the Nation's Health* 24 (April 1934):401–2; Lang, "A Most Valuable Accident," 58–61; "Radium Victim No. 41," *Life* 31 (December 17, 1951): 81.
28. Schubert and Lapp, *Radiation,* 112.
29. "Deadly Radium Gas," *Literary Digest* 101 (June 15, 1929): 19.
30. In 1904 Pierre Curie himself had exposed animals to various concentrations of radon gas, and the animals died within a few hours. Schubert and Lapp, *Radiation,* 111.
31. S. C. Lind, "Radium Production in America," *Chemical and Metallurgical Engineering* 26 (May 31, 1922): 1012.

Chapter 2

1. Edward Teller and Alan Brown, *The Legacy of Hiroshima* (New York: Doubleday, 1962).
2. Henry DeWolf Smyth, *Atomic Energy for Military Purposes: The Official Report on the Development of the Atomic Bomb under the Auspices of the United States Government, 1940–1945* (Princeton: Princeton University Press, 1945), 226.

3. Corbin Allardice and Edward R. Trapnell, *The Atomic Energy Commission* (New York: Praeger, 1974), 26. The best review of the nation's early nuclear consciousness is Paul Boyer, *By the Bomb's Early Light: American Thought and Culture at the Dawn of the Atomic Age* (New York: Pantheon Books, 1985).
4. Atomic Energy Act, 60 Stat. 775 (1946). James R. Newman and Byron S. Miller, *Control of Atomic Energy: A Study of Its Social, Economic, and Political Implications* (New York: McGraw-Hill, 1948), 7–13. See generally Richard G. Hewlett and Oscar E. Anderson, Jr., *The New World, 1939–1946: A History of the United States Atomic Energy Commission* (University Park: Pennsylvania State University Press, 1962), chaps. 12–14; 92 *Congressional Record* 6097 (1946); U.S. Congress, Joint Committee on Atomic Energy, *Hearings Before the Joint Committee on Atomic Energy on S. 3323 and H.R. 8862*, 83d Cong., 2d sess. (1954), 291.
5. Harold P. Green and Alan Rosenthal, *Government of the Atom* (New York: Atherton Press, 1963), 2–3; Steven L. Del Sesto, *Science, Politics, and Controversy: Civilian Nuclear Power in the United States, 1946–1974* (Boulder, Colo.: Westview Press, 1979), 14–17. Section 4(e) of the Atomic Energy Act provides that nobody shall produce, transfer, or acquire any facilities for the production of fissionable material without an AEC license. Section 5(b) was a similar provision limiting the transfer of source material. Section 5(b)(1) states that the term "source material" means "uranium, thorium, or any other material which is determined by the Commission, with the approval of the President, to be particularly essential to the production of fissionable materials; but includes ores only if they contain one or more of the foregoing materials in such concentration as the Commission may by regulation determine from time to time." The AEC regulations further define "source material" as "(1) uranium or thorium, or any combination thereof, in any physical or chemical form or (2) ores which contain by weight one twentieth of one percent (0.05%) of (i) uranium, (ii) thorium, or (iii) any combination thereof." 10 CFR Sec. 40.4(h). Section 7 prohibits a person from producing or exporting any device utilizing fissionable material or atomic energy or from utilizing fissionable material or atomic energy with or without such a device, unless licensed by the AEC. U.S. Atomic Energy Commission, *Letter from the Chairman and Members of the United States Atomic Energy Commission (Second Semiannual Report of the United States Atomic Energy Commission)* (Washington, D.C.: Government Printing Office, 1947), 17.
6. Byron S. Miller, "A Law Is Passed—The Atomic Energy Act of 1946," *University of Chicago Law Review* 15 (Summer 1948): 799–829; Green and Rosenthal, *Government of the Atom*, 25–30.
7. See William Gamson and Andre Modigliani, "Media Discourse and Public Opinion on Nuclear Power," *American Journal of Sociology* 95 (1989): 1–37.
8. Frederick Soddy, *Interpretation of Radium* (London: John Murray, 1909), 229, 244.
9. H. G. Wells, *The World Set Free: A Story of Mankind* (London: Macmillan, 1914); "When H. G. Wells Split the Atom: A 1914 Preview of 1945," *The Nation* 161 (August 18, 1945): 154–56. Leo Szilard claimed that his 1932 reading of *The World Set Free*, combined with his experience with nuclear physics, led him to believe in the military potential of fission weapons. W. Warren Wagar, "Toward a World Set Free: The Vision of H. G. Wells," *Futurist* 17 (1983): 24–31.
10. Stephen Hilgartner, Richard C. Bell, and Rory O'Connor, *Nukespeak: Nuclear Language, Visions, and Mindset* (San Francisco: Sierra Club Books, 1982), xiii–xiv.
11. Ernest Rutherford and Frederick Soddy, "Radioactive Change," *London, Edinburgh, and Dublin Philosophical Magazine and Journal of Science* 6 (1903): 590; Alfred Romer,

ed., *The Discovery of Radioactivity and Transmutation* (New York: Dover, 1964), 165–66.

12. For example, the chemical reaction of burning one kilogram of coal provides 8.5 kilowatt-hours of electricity. But one kilogram of coal would yield 25 billion kilowatt-hours if it could be converted to energy through nuclear reaction. Smyth, *Atomic Energy for Military Purposes*, 2.

13. "Atomic Energy Cannot Compete as Power Source," *Science News Letter* 35 (April 8, 1939): 217. Before World War II, Alfred Nier, of the University of Minnesota, and Kenneth Kingdom and H. C. Pollock, both of General Electric, isolated a uranium isotope (235) through a costly and slow process involving the use of a mass spectronometer. Their yield was one ten-billionth of a pound of uranium isotope per hour. Because about one pound of uranium isotope was necessary for practical atomic power experiments, Nier's method would have yielded a sufficient quantity in eleven thousand centuries! "Atomic Energy in Ten Years?" *Time* 25 (May 27, 1940): 44–46. Fissionable uranium isotope makes up only about 0.71 percent of natural uranium. In 1939 there was less than one ounce of metallic uranium in the United States. U.S. Department of the Interior, *Uranium Development in the San Juan Basin Region: A Report on Environmental Issues*, Final Report (Fall 1990), I–2, I–9.

14. "Atomic Energy Locked," *Literary Digest* 82 (September 20, 1924): 80; "Infinite Energy Just Out of Reach," *Literary Digest* 83 (November 15, 1924): 26–27; *New York Times*, August 31, 1933.

15. "Atomic Power in Ten Years?" 44–46; Smyth, *Atomic Energy for Military Purposes*, 1–30, 45–87; Arthur H. Compton, *Atomic Quest: A Personal Narrative* (New York: Oxford University Press, 1956), 1–63; John J. O'Neill, "Enter Atomic Power," *Harper's Magazine* 181 (June 1940): 1–10. One piece of uranium reactor fuel the size of a golf ball contains as much energy as 168,000 gallons (4,000 barrels) of oil. One pound of uranium contains the energy potential of 3.3 million pounds (1,650 tons) of coal. U.S. Department of the Interior, *Uranium Development in the San Juan Basin*, I–2.

16. Hewlett and Anderson, *The New World*, 9–14.

17. R. M. Langer, "Fast New World," *Collier's* 106 (July 6, 1940): 18.

18. Ibid., 18–19, 54–55; Jack DeMent and H. C. Drake, *Uranium and Atomic Power* (Brooklyn, N.Y.: Chemical Publishing Company, 1941), 22; O'Neill, "Enter Atomic Power," 1–10; "Atomic Energy in Ten Years?" 44–46.

19. John Hersey, *Hiroshima* (New York: Alfred A. Knopf, 1946).

20. Louis Gwin, *Speak No Evil: The Promotional Heritage of Nuclear Risk Communication* (New York: Praeger, 1990), 29–30, 38–41, 71–83; Christoph Hohenemser, Roger Kasperson, and Robert Kates, "The Distrust of Nuclear Power," *Science* 196 (April 1, 1977): 25–34; Philip Wylie, "Deliverance or Doom," *Collier's* 116 (September 29, 1945): 18–80.

21. Daniel Ford, *The Cult of the Atom: The Secret Papers of the Atomic Energy Commission* (New York: Simon and Schuster, 1982), 24.

22. David E. Lilienthal, *Change, Hope and the Bomb* (Princeton: Princeton University Press, 1963), 63–64.

23. "For the Future," *Newsweek* (August 20, 1945): 59–60; "What Is the Atom's Industrial Future?" *Business Week* (March 8, 1947): 22.

24. Lilienthal, *Change, Hope and the Bomb*, 63.

25. Ibid., 109.

26. Oppenheimer once confessed to President Truman in 1948, "I have blood on my

hands." Peter Goodchild, *J. Robert Oppenheimer: "Shatterer of Worlds"* (London: BBC, 1980), 174, 176.
27. J. Robert Oppenheimer, "The Atom Bomb as a Great Force for Peace," *New York Times Magazine* (June 9, 1946): 7, 59–60.
28. Richard G. Hewlett and Francis Duncan, *Atomic Shield, 1947–1952: A History of the United States Atomic Energy Commission* (University Park: Pennsylvania State University Press, 1969), 99.
29. Leslie M. Groves, *Now It Can Be Told: The Story of the Manhattan Project* (New York: Da Capo Press, 1963), 38.
30. Brian Balogh, *Chain Reaction: Expert Debate and Public Participation in American Commercial Nuclear Power, 1945–1975* (Cambridge: Cambridge University Press, 1991), 80–81.
31. Harold Orlans, *Contracting for Atoms* (Washington, D.C.: Brookings Institution Press, 1967), 1.
32. Hewlett and Duncan, *Atomic Shield*, 99–101.
33. Miller, "A Law Is Passed," 821.
34. Frank G. Dawson, *Nuclear Power: Development and Management of a Technology* (Seattle: University of Washington Press, 1976), 21–22.
35. Balogh, *Chain Reaction*, 80; David E. Lilienthal, *The Journals of David Lilienthal*, vol. 2, *Atomic Energy Years, 1945–1950* (New York: Harper and Row, 1964), 229.
36. Hewlett and Duncan, *Atomic Shield*, 362–64.
37. James R. Newman, "The Atomic Energy Industry: An Experiment in Hybridization," *Yale Law Journal* 60 (December 1951): 1324–25; Green and Rosenthal, *Government of the Atom*, 12; Dawson, *Nuclear Power*, 58–60; Mark Hertsgaard, *Nuclear, Inc.: The Men and Money Behind the Nuclear Power Industry* (New York: Pantheon Books, 1983), 26.
38. U.S. Congress, Joint Committee on Atomic Energy, *Hearings Before the JCAE on Atomic Power Development and Private Enterprise*, 83d Cong., 1st sess. (1953).
39. U.S. Congress, JCAE, *Atomic Power and Private Enterprise*, Joint Committee Print, JCAE, December 1952, 325.
40. U.S. Congress, JCAE, *Atomic Power Development and Private Enterprise*, 2.
41. Ibid., 1–33, 436.
42. Ford, *The Cult of the Atom*, 59. The most resounding economic condemnation of commercial atomic power was Sam H. Schurr and Jacob Marshank, *Economic Aspects of Atomic Power* (Princeton: Princeton University Press, 1950).
43. For examples of arguments made in support of commercial atomic development, see generally U.S. Congress, Joint Committee on Atomic Energy, *Atomic Development and Private Enterprise: Hearings Before the Joint Committee on Atomic Energy*, 81st Cong., 1st sess., May 26, 1949.
44. Atomic Energy Act of 1954, 68 Stat. 919 (1954).
45. Richard S. Lewis, *The Nuclear-Power Rebellion* (New York: Viking Press, 1972), 26.
46. Philip Mullenbach, *Civilian Nuclear Power* (New York: Twentieth Century Fund, 1963), 3–7. Some observers embraced the active participation of the government in development of atomic energy. "What has happened," asked James Newman, "to the innocent notion that the benefits of atomic energy should accrue to the nation as a whole—without the prior drain of private profits—since the resource itself was brought to fruition by public funds? Somewhere along the circular route of national policy this point got lost." Newman, "The Atomic Energy Industry," 1391.
47. David F. Cavers, "The Atomic Energy Act of 1954," *Scientific American* 191 (November 1954): 35.

48. P.L. 85-256 (71 Stat. 576) (1957). The Price-Anderson Act was an amendment to the Atomic Energy Act of 1954. In 1957 the AEC issued a report produced by its Brookhaven National Laboratory that outlined the consequences of an accident at a nuclear power plant. The report estimated that a serious accident near a large city would result in 3,400 deaths, 43,000 injuries, and $7 billion in property damage. A General Electric representative testified during hearings on the proposed Price-Anderson legislation that the company could not move forward with its atomic energy program "with a cloud of bankruptcy hanging over its head." Recognizing that no industry or private insurer could ever cover such vast losses, the Price-Anderson Act capped liability for any nuclear accident at $560 million. That amount was derived arbitrarily by adding the total private insurance the industry could raise, $60 million, with the government guaranteeing that it would pay out the remaining $500 million if necessary. The act also required industry to pay into an insurance pool that would (and did) eventually cover the amount guaranteed by the government. Finally, the act stipulated that Congress could authorize higher amounts of compensation if damages exceeded $560 million. Michelle Adato, James MacKenzie, Robert Pollard, and Ellyn Weiss, *Safety Second: The NRC and America's Nuclear Power Plants* (Bloomington: Indiana University Press, 1987), 3; Dawson, *Nuclear Power*, 123, 129.
49. The original version of the 1954 act contained twenty-five references to public health and safety.
50. Green and Rosenthal, *Government of the Atom*, 29–30.
51. Ibid., 247–52.
52. Ibid., 6–20, 247–52; Morgan Thomas, *Atomic Energy and Congress* (Ann Arbor: University of Michigan Press, 1956), 23–140. The symbiotic relationship between the AEC and the JCAE eroded around 1955 when disagreements arose between the Republican administration and the Democrat-dominated JCAE. In general, the JCAE demanded that the AEC and the executive branch adopt policies that it favored and used its oversight authority to compel the AEC to comply with JCAE objectives. By 1957 the JCAE exercised enormous influence over the AEC.
53. James M. Jasper, "Nuclear Policy as Projection: How Policy Choices Can Create Their Own Justification," in John Byrne and Steven M. Hoffman, eds., *Governing the Atom: The Politics of Risk* (New Brunswick, N.J.: Transaction, 1996), 49.
54. Ibid., 50.
55. Gerald H. Clarfield and William M. Wiecek, *Nuclear America: Military and Civilian Nuclear Power in the United States 1940–1980* (New York: Harper and Row, 1984), 273. When the Shippingport reactor went on line in 1957, it produced electricity at about ten times the cost of a coal-fired electrical plant. The AEC's Brookhaven Laboratory theorized that an atomic accident would kill 3,000 people immediately and injure 40,000 more. U. S. Atomic Energy Commission, *Theoretical Possibilities and Consequences of Major Accidents in Large Nuclear Power Plants* (Washington, D.C.: Government Printing Office, 1958). See generally James M. Jasper, *Nuclear Politics: Energy and the State in the United States, Sweden, and France* (Princeton: Princeton University Press, 1990); and Jasper, "Nuclear Policy as Projection."
56. Richard O. Niehoff, "Organization and Administration of the United States Atomic Energy Commission," *Public Administration Review* 8 (May 1948): 102.

Chapter 3

1. Jerome Strauss, "Radium, Uranium, and Vanadium," in G. A. Roush, ed., *The Mineral Industry: Its Statistics, Technology, and Trade during 1939* (New York: McGraw-Hill,

1940), 520; Jerome Strauss, "Radium, Uranium, and Vanadium," in G. A. Roush, ed., *The Mineral Industry: Its Statistics, Technology, and Trade during 1940* (New York: McGraw-Hill, 1941), 542; Jack DeMent and H. C. Drake, *Uranium and Atomic Power* (Brooklyn, N.Y.: Chemical Publishing Co., 1945), 21–22; Richard G. Hewlett and Oscar E. Anderson, *The New World, 1939–1946: A History of the United States Atomic Energy Commission* (University Park: Pennsylvania State University Press, 1962), 46–47.
2. Leonard J. Arrington and Anthony T. Cluff, *Federally Financed Industrial Plants Constructed in Utah during World War II* (Logan: Utah State University Press, 1969); Gary Lee Shumway, "A History of the Uranium Industry on the Colorado Plateau" (Ph.D. diss., University of Southern California, 1970), 101–20; *Durango Herald-Democrat*, June 17, 1948.
3. William L. Laurence, "The Atom Gives Up," *Saturday Evening Post* (September 7, 1940): 62.
4. Hewlett and Anderson, *The New World*, 115; Jesse C. Johnson, "The Romance of Uranium Mining," *Science Digest* 40 (September 1956): 59.
5. Richard G. Hewlett and Francis Duncan, *Atomic Shield, 1947–1952: A History of the United States Atomic Energy Commission* (University Park: Pennsylvania State University Press, 1969), 148. In 1938 Americans produced only about twenty-six tons of uranium oxides and salts, an insignificant amount compared to what the atomic program required. Hewlett and Anderson, *The New World*, 26.
6. Benjamin N. Webber, "Geology and Ore Resources of the Uranium-Vanadium Depositional Province of the Colorado Plateau Region," Union Mines Development Corporation, *Final Report* (January 1947): 3. Webber's unpublished manuscript is cited in Shumway, "A History of the Uranium Industry on the Colorado Plateau," 159–62; Charles H. Viol, "Radium Production," *Science* 49 (March 7, 1919): 227; Richard B. Moore, "Radium Production," *Science* 49 (June 13, 1919): 564–66; Harold B. Meyers, "The Great Uranium Glut," *Fortune* 69 (February 1964): 111.
7. "Uranium Is Too Scarce for Use as Source of Fuel," *Science News Letter* 53 (May 1, 1948): 283–84. AEC chairman, David E. Lilienthal, denied Millikan's assessment, noting that there was "plenty of uranium in the world for the atomic age." "Plenty of Uranium," *Science News Letter* 54 (December 25, 1948): 403.
8. James R. Newman and Byron S. Miller, *Control of Atomic Energy: A Study of Its Social, Economic, and Political Implications* (New York: McGraw-Hill, 1948), 97–98.
9. Hewlett and Duncan, *Atomic Shield*, 47–48, 147–49, 173; Hewlett and Anderson, *The New World*, 285–88, 654; Meyers, "The Great Uranium Glut," 111; Shumway, "A History of the Uranium Industry on the Colorado Plateau," 162.
10. Hewlett and Duncan, *Atomic Shield*, 147–48; "Pure Science," *Time* (August 11, 1947): 25. For an overview of America's foreign uranium acquisition program, see Jonathan E. Helmreich, *Gathering Rare Ores: The Diplomacy of Uranium Acquisition, 1943–1954* (Princeton: Princeton University Press, 1986).
11. Meyers, "The Great Uranium Glut," 111.
12. "How to Find Uranium," *Time* (April 21, 1947): 86.
13. David E. Lilienthal, "Statement for press conference in Denver, Colorado, December 17, 1948," published in U.S. Congress, Joint Committee on Atomic Energy, *Atomic Power and Private Enterprise*, Joint Committee Print, JCAE, December 1952.
14. Hewlett and Anderson, *The New World*, 425–26.
15. U.S. Congress, Senate, *Senate Report No. 1211*, 79th Cong., 2d sess., 1946, 19.
16. U.S. Atomic Energy Commission, *Letter from the Chairman and Members of the United*

States Atomic Energy Commission (Third Semiannual Report of the United States Atomic Energy Commission) (Washington, D.C.: Government Printing Office, 1948), 4.
17. See sec. 5(b)(7) of the Atomic Energy Act of 1946; Newman and Miller, Control of Atomic Energy, 97–105; U.S. Atomic Energy Commission, Fourth Semiannual Report (Washington, D.C.: Government Printing Office, 1948), 50.
18. Robert D. Nininger, Minerals for Atomic Energy (New York: D. Van Nostrand, 1954), 4.
19. Ibid., 5; John K. Gustafson, "Uranium Resources," Scientific Monthly 69 (August 1949): 120; AEC, Third Semiannual Report, 4.
20. AEC, Fourth Semiannual Report, 50.
21. Nininger, Minerals for Atomic Energy, 5–6.
22. "Private Development of Uranium Is Encouraged," Engineering and Mining Journal 149 (June 1948): 67.
23. AEC, Fourth Semiannual Report, 49–50; "U.S. Atomic Energy Commission Announces Program to Stimulate Production of Domestic Uranium," Engineering and Mining Journal 149 (May 1948): 108; Gustafson, "Uranium Resources," 119; Larry L. Meyer, "The Time of Great Fever: U-Boom on the Colorado Plateau," American Heritage 32 (June–July 1981): 74. The AEC "Domestic Uranium Program Circular 5, Revised," was amended in September 1953 to extend the guaranteed uranium oxide price schedule to March 31, 1962, from its previous expiration date of March 31, 1958. The AEC "Uranium Program Circular 6," which established the bonus program for discovery and certain production of uranium ore, was also extended through February 28, 1957, from its previous expiration date of February 28, 1954. Lewis L. Strauss to W. Sterling Cole, September 22, 1953, Box 146, Joint Committee on Atomic Energy General Correspondence File, Record Group 128, National Archives and Records Administration, Washington, D.C. (hereafter JCAE Papers).

The bonus program was a resounding success in stimulating prospecting and mining and was very profitable for the miners. By August 1953 the AEC had made 1,347 individual bonus payments, and by November 1953 it had made 1,626 individual payments totaling $2,535,000. In April 1955 the bonus payments passed the $5 million mark, with an average bonus payment rate of about $195,000 per month. Press and Radio Release nos. 38, 48, 115, August 5, 1953, November 17, 1953, and April 8, 1955, Box 353, JCAE Papers. "Private Development of Uranium Is Encouraged," Engineering and Mining Journal 149 (June 1948): 67; "AEC Increases Price for Uranium Bearing Ores," Engineering and Mining Journal 149 (June 1948): 103; U.S. Atomic Energy Commission, Atomic Energy and the Life Sciences (Sixth Semiannual Report) (Washington, D.C.: Government Printing Office, 1949), 3; U.S. Atomic Energy Commission, Seventh Semiannual Report of the Atomic Energy Commission (Washington, D.C.: Government Printing Office, 1950), 3.
24. Shumway, "A History of the Uranium Industry on the Colorado Plateau," 176–78.
25. Hewlett and Duncan, Atomic Shield, 173.
26. Meyer, "The Time of Great Fever," 74.
27. U.S. Atomic Energy Commission, Seventh Semiannual Report of the Atomic Energy Commission (Washington, D.C.: Government Printing Office, 1950), 4; "Uranium Grows Up—Big Business Now," U.S. News and World Report (April 6, 1956): 91.
28. "Colorado Plateau Uranium Population Doubles in 2 Years," Denver Post, February 6, 1955; "History's Greatest Metal Hunt," Life 38 (May 23, 1955): 25–35; Thomas E. Gillingham, "Uranium," Mining Congress Journal 40 (February 1954): 116–18; "Uranium, 1956," True West 3 (May–June 1956): 4–7, 35. Estimates suggest that the population of the Colorado Plateau doubled in two years. The population of the

company town of Uravan grew from 800 in 1952 to 1,500 in 1955; Naturita went from 39 to 1,200 people. "Adjusting to A-Boom Easy for Company-Owned Towns," *Denver Post*, November 17, 1954; "Uranium Industry Boosts Naturita Population," *Denver Post*, November 16, 1954. See generally Michael A. Amundson, "Home on the Range No More: The Boom and Bust of a Wyoming Uranium Mining Town, 1957–1988," *Western Historical Quarterly* 26 (1995): 483–505; Stephen I. Schwartz, *Atomic Audit: The Costs and Consequences of U.S. Nuclear Weapons since 1940* (Washington, D.C.: Brookings Institution Press, 1998).

29. Arthur R. Gomez, *Quest for the Golden Circle: The Four Corners and the Metropolitan West, 1945–1970* (Albuquerque: University of New Mexico Press, 1994), 26; Holger Albrethsen, Jr., and Frank E. McGinley, *Summary History of Domestic Uranium Procurement under U.S. Atomic Energy Commission Contracts* (Grand Junction, Colo.: U.S. DOE Area Field Office, 1982), B-2-3; Raymond W. Taylor and Samuel W. Taylor, *Uranium Fever, or No Talk under $1 Million* (New York: Macmillan, 1970), 249–53.

30. "Pennies for Uranium," *Time* 63 (April 5, 1954): 89.

31. "Why Buy Uranium Stocks Now?" *Newsweek* 46 (August 29, 1955): 57–58.

32. "Colorado Plateau: Fabulous Treasure House of Energy," Moab, Utah, *Times-Independent*, Special Edition, 1954; U.S. Congress, House Committee on Interstate and Foreign Commerce, *Hearings on Amendments to Securities Act of 1933 (Hearings on H.R. 5701 and H.R. 9319)*, 84th Cong., 1st and 2d sess., July 20, 1955–May 9, 1956; Perrin Stryker, "The Great Uranium Rush," *Fortune* 50 (August 1954): 89–93, 148–58; "The Future of Uranium," *Time* (February 14, 1955): 94; "SEC Wars on Uranium, Oil Promotions," *Business Week* (July 23, 1955): 66; "Punctured Boom," *Business Week* (December 17, 1955): 124–26; Meyer, "The Time of Great Fever," 74–80. For an overview of the uranium boom, see Raye C. Ringholz, *Uranium Frenzy: Boom and Bust on the Colorado Plateau* (Albuquerque: University of New Mexico Press, 1991).

33. Vernon Pick, who had no previous mining experience, found and sold a uranium mine for $9 million and an airplane. Steen, a down-and-out Texas geologist, found that his Mi Vida mine had at least $60 million in proven reserves. Dysart was an oil "wildcatter" and con artist who sold stock in usually failing oil companies, had been run out of Utah by the state Securities and Exchange Commission, and had served sixteen months in a Los Angeles jail for securities fraud. When she was seventy-seven years old, Mid-Continent Exploration Company struck uranium on her property at Ambrosia Lake, New Mexico. The seventeen-foot-thick layer of uranium contained more than 65 percent of all estimated U.S. uranium reserves as of 1957. Dysart's 17.5 percent royalty checks made her an estimated $100 million fortune by 1964. See generally Ringholz, *Uranium Frenzy*; Maxine Newell, *Charlie Steen's Mi Vida* (Moab, Utah: Moab's Printing Place, 1992); Johnson, "The Romance of Uranium Mining," 58–62; Meyer, "The Time of Great Fever," 74–79; *Time* (September 30, 1957): 89; *Time* (June 27, 1955): 80; Burt Meyers, "Uranium Jackpot," *Engineering and Mining Journal* 154 (September 1953): 72–75; Elizabeth Pope, "The Richest Town in the U.S.A.," *McCall's* (December 1956): 38–39, 99–104; *Business Week* (August 1, 1953): 28; *Denver Post*, November 29, 1964; U.S. Department of the Interior, *Uranium Development in the San Juan Basin Region*, 1–10.

34. The economists Sam H. Schurr and Jacob Marschak made their appraisal of the future of atomic energy in 1950. Far from the early prophecies of limitless and free energy, they made surprisingly accurate predictions. They forecasted that, based on 1946 prices, atomic energy would initially cost nearly twice as much as conventionally pro-

duced electricity and that at best it would cost only slightly less than fossil fuel–generated power. Sam H. Schurr and Jacob Marschak, *Economic Aspects of Atomic Power* (Princeton: Princeton University Press, 1950); *Newsweek* 36 (November 27, 1950): 72–73.

35. David S. Teeple, "The Coming Uranium Bust," *American Mercury* 81 (September 1955): 34–39; "Punctured Boom," *Business Week* (December 17, 1955): 124.
36. "Coming of the Giants," *Time* (May 28, 1956): 91.
37. Ibid.
38. *Business Week* (April 21, 1956): 30–32; "Uranium Grows Up—A Big Business Now," *U.S. News and World Report* 40 (April 6, 1956): 90–92; *Time* 55 (February 14, 1955): 94–96. Between 1947 and 1970 the AEC purchased 174,000 tons of uranium oxide from domestic sources, which represented almost 55 percent of its total acquisitions for that period. 73,000 tons were purchased from Canada, and 67,600 tons came from abroad. Jonathan E. Helmreich, *Gathering Rare Ores: The Diplomacy of Uranium Acquisition, 1943–1954* (Princeton: Princeton University Press, 1986), 226.
39. "Uranium Industry Leaps to Maturity," *Business Week* (April 21, 1956): 30. Economists refer to a market with only one buyer, as was the case with the uranium industry, as a "monopsony." Under these conditions, profits are guaranteed only for the life of the contract. Moreover, until November 24, 1958, the government uranium-buying program purchased all the industry's output, regardless of the amount. Consequently, there was considerable incentive for uranium companies to maximize profits during the contract period, especially after the AEC announced the end of its unlimited purchasing program. Gerald Nash, *The American West in the Twentieth Century: A Short History of an Urban Oasis* (Englewood Cliffs, N.J.: Prentice-Hall, 1973), 229–35.
40. For the most comprehensive discussion of milling technology, see Robert C. Merritt, *The Extractive Metallurgy of Uranium* (Golden: Colorado School of Mines Research Institute, 1971).
41. Herman Fleck, "A Series of Treatises on the Rare Metals," *Proceedings of the Colorado Scientific Society*, vol. 11 (Denver: Colorado Scientific Society, 1916), 156–59; J. M. Boutwell, "Vanadium and Uranium in Southeastern Utah," *U.S. Geological Survey Bulletin* 260 (Washington, D.C.: Government Printing Office, 1905), 200–210; Don Sorensen, "Wonder Mineral: Utah's Uranium," *Utah Historical Quarterly* 31 (Summer 1963): 282–83; U.S. Geological Survey, *Mineral Resources of the United States, 1901* (Washington, D.C.: Government Printing Office, 1902), 270; U.S. Geological Survey, *Mineral Resources of the United States, 1906*, 526; Richard B. Moore and Karl L. Kithil, "A Preliminary Report on Uranium, Radium, and Vanadium," *U.S. Bureau of Mines Bulletin* 70 (Washington, D.C.: Government Printing Office, 1913), 9–29; Kathleen Bruyn, *Uranium Country* (Boulder: University of Colorado Press, 1955), 49–50; U.S. Department of Interior, National Park Service, *Vanadium Corporation of America Naturita Mill*, Historic American Engineering Record, Report HAER No. CO-81, 6.
42. HAER No. CO-81, 8; Richard P. Fischer, J. C. Haff, and J. F. Rominger, "Vanadium Deposits Near Placerville, San Miguel County, Colorado," *Colorado Scientific Society Proceedings* 15 (1947): 119–20; "Highlights in the History of Vanadium," *Mining and Contracting Review* 42 (July 31, 1942): 12–13.
43. U.S. Department of Interior, *Minerals Yearbook, 1940* (Washington, D.C.: Government Printing Office, 1941), 626; Shumway, "A History of the Uranium Industry on the Colorado Plateau," 101–2; HAER No. CO-81, 8. The most common

practice was for the large companies, whether they needed the ore or not, to lease or buy the claims that supplied the small mills with raw material and to process the ore at their own mills, thereby depriving the small mills of their customers. Shumway, "A History of the Uranium Industry on the Colorado Plateau," 110–13.

44. *Minerals Yearbook, 1940*, 625; U.S. Department of Interior, *Minerals Yearbook, 1941* (Washington, D.C.: Government Printing Office, 1942), 637; U.S. Department of Interior, *Minerals Yearbook, 1942* (Washington, D.C.: Government Printing Office, 1943), 665–67; Shumway, "A History of the Uranium Industry on the Colorado Plateau," 123–25; HAER No. CO-81, 9; James Keener and Christine Bebee Keener, *Colorado Highway 141, Unaweep to Uravan: Travel through 1.7 Billion Years into the Atomic Age* (Grand Junction, Colo.: Grand River Publishing, 1988), 40. The dubious business practices of VCA and Union Carbide's USVC, including evidence of price fixing, competition elimination, and market monopoly, convinced U.S. Attorney General Thomas Clark to initiate a criminal suit against officials of those companies for violating the Sherman Anti-Trust Act. The criminal action failed and the case ended in 1948. A subsequent criminal action on reduced charges also failed. A civil action against the companies succeeded in 1959, resulting in triple damages of nearly $5 million to the independent miners. "Elemental Strife," *Business Week* (September 1, 1945): 101–2; "Vanadium Quiz," *Business Week* (July 14, 1945): 36–38; Shumway, "A History of the Uranium Industry on the Colorado Plateau," 137–38, 174–75.
45. Hewlett and Anderson, *The New World*, 291–94.
46. Albrethsen and McGinley, *Summary History*, 4.
47. Ibid., 6, B-7; "New Mills Break Up Uranium Bottleneck," *Business Week* (August 6, 1955): 102–3, 105.
48. "New Mills Break Up Uranium Bottleneck," *Business Week* (August 6, 1955): 102.
49. Ibid., 103; Amundson, "Home on the Range No More," 486–87.
50. K. D. Nichols to Corbin Allardice, December 20, 1954, Box 146, JCAE Papers. See generally Albrethsen and McGinley, *Summary History*.
51. "New Mills Break Up Uranium Bottleneck," *Business Week* (August 6, 1955): 102–3. The Moab mill was built by Uranium Reduction Company and controlled by Charles Steen's Utex Exploration Company. Such diverse financial interests as Chemical Corn Exchange Bank and New York Life Insurance Company, the first life insurance company to finance a uranium mill, backed Steen. Kerr-McGee Oil Industries, Inc., of Oklahoma City (the Kerr was Senator Robert Kerr of Oklahoma) financed the mill at Shiprock, New Mexico, marking the first time that an oil company built and operated a uranium milling facility; *Time* 65 (June 27, 1955): 80; Frank G. Dawson, *Nuclear Power: Development and Management of a Technology* (Seattle: University of Washington Press, 1976), 161–64; "AEC Breaks Uranium Log Jam," *Business Week* (October 11, 1958): 27; *Business Week* (April 21, 1956): 32.
52. John Byrne and Steven M. Hoffman, "The Ideology of Progress and the Globalization of Nuclear Power," in John Byrne and Steven M. Hoffman, eds., *Governing the Atom: The Politics of Risk* (New Brunswick, N.J.: Transaction, 1996), 11.

Chapter 4

1. Elizabeth S. Rolph, *Nuclear Power and Public Safety: A Study in Regulation* (Lexington, Mass.: D. C. Heath, 1979), 22, 26–29. See generally Louis Gwin, *Speak No Evil: The Promotional Heritage of Nuclear Risk Communication* (New York: Praeger, 1990); "Putting the Atom to Work," *Popular Mechanics* (May 1938): 690; R. M. Langer, "Fast New World," *Collier's* 106 (July 6, 1940): 54.

2. Critics have challenged the AEC's industrial safety claims. The AEC, they allege, gave itself good marks on safety because it focused only on lost-time accidents. The Commission ignored radiation contamination of workers, the most serious but least obvious on-the-job injuries at nuclear facilities. H. Peter Metzger, *The Atomic Establishment* (New York: Simon and Schuster, 1972), 116–17.
3. U.S. Atomic Energy Commission, *Control of Radiation Hazards in the Atomic Energy Program (Eighth Semiannual Report)* (Washington, D.C.: Government Printing Office, 1950), 20.
4. Herbert A. Simon, *Reason in Human Affairs* (Stanford: Stanford University Press, 1983), 96–97.
5. Joseph J. Katz and Eugene Rabinowitch, *The Chemistry of Uranium (Part 1): The Element, Its Binary and Related Compounds* (New York: McGraw-Hill, 1951), 111–32; Thomas C. Hollocher and James J. MacKenzie, "Radiation Hazards Associated with Uranium Mill Operations," in Union of Concerned Scientists, *The Nuclear Fuel Cycle* (Cambridge, Mass.: MIT Press, 1975), 45; U.S. Congress, Subcommittee on Air and Water Pollution of the Senate Committee on Public Works, *Radioactive Water Pollution in the Colorado River Basin*, 89th Cong., 2d sess., May 6, 1966, 4–6, 11 (hereafter *Colorado River Basin Hearing*).

  The most dramatic tailings dam failure occurred at the Kerr-McGee Uranium Mill at Shiprock, New Mexico, on August 22–23, 1960, in which mill by-product was released into the San Juan River. Mill officials did not immediately report the dam break and river contamination to local, state, or federal officials. Not until after a local newspaper had reported on the dead fish in the river, approximately five days after the spill had occurred, did Kerr-McGee finally acknowledge the incident and report it to the New Mexico and Utah state governments, the appropriate federal government agencies (including the AEC Licensing and Regulation Division), and the Public Health Service Indian Health Hospital at Shiprock. Fortunately, the water treatment plant for Shiprock, which usually relied on the San Juan River water drawn from its collection facility immediately below the mill, was at the time of the spill drawing water from an alternate source. Initial reports indicated that about 250,000 gallons of liquid had been released, but a revised estimate placed the amount at between 470,000 and 780,000 gallons. The liquid contained radioactive materials and toxic organic contaminants. For a brief time, radioactive materials in the spilled waste raised the amount of radiation in the river to about twenty times the permissible level set forth in the USPHS's "Drinking Water Standards." Further analysis of mud from the channel into which the spilled waste flowed showed high radioactivity that was not the result of the spill alone but more likely reflected long-term seepage from the tailings impoundment ponds. In its report on the spill, the USPHS concluded that it was unlikely anyone suffered dangerous radiation exposure from the spill and that the spill had resulted from the poor construction of the ponds or the fact that the ponds were filled beyond capacity, or both. U.S. Department of Health, Education and Welfare, Public Health Service (Region VIII, Denver, Colorado), *Shiprock, New Mexico Uranium Mill Accident of August 22, 1960*, Colorado River Basin Water Quality Control Project (January 1963), located in the archives of the New Mexico Environmental Department, Hazardous Waste Bureau, Mixed Waste Section, Santa Fe, N.Mex. (hereafter N.Mex. Environmental Department).
6. *Colorado River Basin Hearing*, 4–6.
7. Ibid., 11; Hollocher and MacKenzie, "Radiation Hazards Associated with Uranium Mill Operations," 47; Metzger, *The Atomic Establishment*, 162, citing U.S. Atomic

Energy Commission, "Uranium Mill Tailings," presentation for an "AEC Pollution Meeting" by the Division of Operational Safety, AEC (September 1967).
8. "Transcript of Conference on Interstate Pollution of the Animas River, Colorado-New Mexico," Santa Fe, N.Mex., April 29, 1958, 11–13, 36–40 (hereafter Animas River Transcript), citing R. F. Poston, *Uranium Milling Waste Studies–Colorado and Utah*, unpublished memorandum, Western Gulf and Colorado Basin Office, USPHS (March 19, 1951). A copy of the Animas River Transcript is located in the N.Mex. Environmental Department. E. C. Tsivoglou, A. F. Bartsch, D. L. Rushing, and D. A. Holiday, *Report of Survey of Contamination of Surface Waters by Uranium Recovery Plants*, USPHS, Robert A. Taft Sanitary Engineering Center, Cincinnati, Ohio, 1956.
9. Aleck Alexander, memorandum to Division of Sanitary Engineering Services et al., February 20, 1958, Box 4, U.S. Public Health Service Accession No. 90-62A-672, National Archives and Records Administration, Washington, D.C. (hereafter PHS Papers); Colorado Department of Public Health, *Uranium Wastes and Colorado's Environment*, 2d ed. (Denver: Colorado Department of Public Health, 1971), 10, citing S. D. Shearer et al., *Waste Characteristics for the Acid-Leach Solvent Extraction Uranium Refining Process, I, Gunnison Mining Company*, USPHS Technical Report W62–17 (Robert A. Taft Sanitary Engineering Center, Cincinnati, Ohio, 1962); J. B. Cohen et al., *Waste Characteristics for the Acid-Leach Solvent Extraction Uranium Refining Process, II, Climax Uranium Company*, USPHS Technical Report W62–17 (Robert A. Taft Sanitary Engineering Center, Cincinnati, Ohio, 1962); J. B. Cohen et al., *Waste Characteristics for the Carbonate-Leach Uranium Extraction Process, I, Homestake–New Mexico Partners Company*, USPHS Technical Report W62–17 (Robert A. Taft Sanitary Engineering Center, Cincinnati, Ohio, 1962); J. R. Pahren et al., *Waste Characteristics for the Carbonate-Leach Uranium Extraction Process, II, Homestake-Sapin Company*, USPHS Technical Report W62–17 (Robert A. Taft Sanitary Engineering Center, Cincinnati, Ohio, 1962); Murry Stein to James Harlan, June 22, 1959, Box 1, Accession No. 90-62A-121, PHS Papers; *Colorado River Basin Hearing*, 3; Hollocher and MacKenzie, "Radiation Hazards Associated with Uranium Mill Operations," 47. Although it is true that radiation levels below the Durango mill were high, the organic-raffinate compounds from the mill process contributed significantly to the unhealthy river conditions. When the mill stopped dumping toxic organic compounds, the Animas began to recover and fish returned to its previously dead reaches.
10. Howard Ball, *Justice Downwind: America's Atomic Testing Program in the 1950s* (Oxford: Oxford University Press, 1986), 39–41.
11. L. K. Olson, memorandum to Harold L. Price, April 15, 1960, Accession No. 9210120199, U.S. Nuclear Regulatory Commission Public Documents Reading Room, Washington, D.C. (hereafter NRC Papers); Neil D. Naiden, memorandum to Harold L. Price, December 7, 1960, Accession No. 9210130222, NRC Papers; AEC, "Proposed Revision of 10 CFR, Part 40, 'Licensing of Source Material' (Note by the Acting Secretary)," AEC-R 18/2 (July 22, 1960), Accession No. 9210120308, NRC Papers. See generally Joseph F. Hennessey, memorandum to Glenn Seaborg et al., June 9, 1966, Box 184, U.S. Atomic Energy Commission Glenn Seaborg Collection, U.S. Department of Energy Archives, Germantown, Md. (hereafter AEC Seaborg Papers).
12. Although this jurisdictional impasse was resolved somewhat in 1959 when Congress amended the 1954 act to clarify state responsibilities, the AEC never really abdicated its central authority over nuclear energy. George T. Mazuzan and J. Samuel Walker, *Controlling the Atom: The Beginnings of Nuclear Regulation, 1946–1962* (Berkeley: University of California Press, 1984), 277–303.

13. Ibid., 309; see generally Howard Ball, *Cancer Factories: America's Tragic Quest for Uranium Self-Sufficiency* (Westport, Conn.: Greenwood Press, 1993), 1–18.
14. *Colorado River Basin Hearing*, 44.
15. Ibid., 46.
16. Grants, N.Mex., *Cibola County Beacon*, April 26, 1958.
17. "AEC Breaks Uranium Log Jam," *Business Week* (October 11, 1958): 27.
18. Lyall Johnson to Harold L. Price, August 24, 1970, Accession No. 9210120432, NRC Papers; 10 CFR Part 20, "Standards for Protection Against Radiation."
19. Ball, *Cancer Factories*, 12.
20. *Colorado River Basin Hearing*, 3; Animas River Transcript, 6, 9; Hollocher and MacKenzie, "Radiation Hazards Associated with Uranium Mill Operations," 47.
21. W. B. Harris et al., *Environmental Hazards Associated with the Milling of Uranium Ore*, HASL-40, U.S. Atomic Energy Commission, Health and Safety Laboratory, New York Operations Office, June 1958, 19, (Table XIII). The twelve mills investigated were Anaconda, Climax, Durango, Edgmont, URC-Moab, Monticello, Naturita, Rifle, Shiprock, Tuba City, Uravan, and Vitro–Salt Lake City.
22. Animas River Transcript, 27, 32; J. Samuel Walker, *Containing the Atom: Nuclear Regulation in a Changing Environment, 1963–1971* (Berkeley: University of California Press, 1992), 258; Mazuzan and Walker, *Controlling the Atom*, 314–15.
23. *Colorado River Basin Hearing*, 11; Animas River Transcript, 4, 26–33.
24. James T. Ramey, memorandum to All Members of Subcommittee on Raw Materials, June 27, 1960, Box 251, Joint Committee on Atomic Energy General Correspondence File (Record Group 128), National Archives and Records Administration, Washington, D.C. (hereafter JCAE Papers)
25. Ibid. Animas River Transcript, 29, 31.
26. Animas River Transcript, 48.
27. John A. McCone, memorandum to President Eisenhower, July 24, 1959, Box 7717, Atomic Energy Commission Secretariat File, United States Department of Energy Archives, Germantown, Md. (hereafter AEC Secretariat Papers); Luther J. Carter, "Uranium Mill Tailings: Congress Addresses a Long-Neglected Problem," *Science* 202 (October 13, 1978): 191–95.
28. *Colorado River Basin Hearing*, 12, 41, 97–98; Arve H. Dahl, memorandum to "the record," February 1, 1960, Box 13, Accession No. 90-65A-533, PHS Papers; Olson to Price, April 15, 1960. The key provisions on which Olson, the AEC's general counsel, based his opinion were sections 62, 161(b), and 161(i)(3) of the Atomic Energy Act of 1954.
29. Arthur L. Warner to Roy L. Cleere and Donald Walker, September 20 and 24, 1963 (AEC-R 18/17, November 20, 1963), Accession No. 9210120224, NRC Papers; Thomas J. Coupe to Gordon Allott, Peter Dominick, and Wayne Aspinall, May 11, 1963, Box 134, Wayne Aspinall Papers, Penrose Library, University of Denver, Denver, Colo. (hereafter Aspinall Papers). See also Colorado Department of Public Health, "Report on Control of Uranium Mill Tailings," Occupational and Radiological Health Section, October 28, 1966, Box 9937, Colorado State Archives, Denver (hereafter Colorado State Archives); John T. Conway, memorandum to Wayne Aspinall, October 3, 1963, Box 144, Aspinall Papers.
30. W. B. McCool, "Note by the Secretary," October 11, 1965 (AEC-R 18/28, October 11, 1965), Accession No. 9210120270, NRC Papers.
31. Ibid., *Colorado River Basin Hearing*, 20.
32. Donald Nussbaumer, memorandum to files, November 14, 1963, Accession No.

9210120284, NRC Papers; Harold L. Price to John T. Conway, October 24, 1963 (AEC-R 18/17, November 20, 1963), Accession No. 9210120224, NRC Papers; Price to Conway, October 24, 1963, Box 144, Aspinall Papers; W. B. McCool to the file, January 7, 1966, Box 7717, AEC Secretariat Papers; McCool, "Note by the Secretary," October 11, 1965; *Colorado River Basin Hearing*, 44; Naiden to Price, December 7, 1960.

33. *Colorado River Basin Hearing*, 20, 41, 45; Colorado Department of Public Health, "Report on Control of Uranium Mill Tailings"; R. E. Hollingsworth and Harold L. Price, memorandum to Glenn Seaborg et al., December 2, 1965 (AEC-R 18/32, December 3, 1965), Accession No. 9210120297, NRC Papers.

34. One consequence of the Animas River Conference was that the FWPCA, which had been instrumental in gathering data about radium water pollution in the Colorado Basin, became involved in enforcement measures designed to reduce radioactive water pollution from uranium mills. As evidence of extensive radium pollution came to light, the FWPCA concluded that radioactive wastes were as serious a health concern as other forms of water pollution and should be dealt with in an equally forceful manner. In 1961 the FWPCA established the Radium Monitoring Network to observe, on a continuous basis, the Colorado River Basin surface water for radium contamination. Dahl to "the record," February 1, 1960; Colorado Department of Public Health, *Uranium Wastes*, 11; *Colorado River Basin Hearing*, 16.

35. U.S. Department of Health, Education and Welfare, *Disposition and Control of Uranium Mill Tailings Piles in the Colorado River Basin* (Denver: Federal Water Pollution Control Administration, Region VII, March, 1966), Box 235, Aspinall Papers, i–ii (hereafter FWPCA Report). The report itself was drafted by the Colorado River Basin Water Quality Control Project of the USPHS. *Colorado River Basin Hearing*, 17, 114–29; Hollingsworth and Price to Seaborg et al., December 2, 1965; James G. Terrill to Robert Lowenstein, August 9, 1963, Box 20, Accession No. 90-66A-484, PHS Papers; Nussbaumer to files, March 14, 1964, Accession No. 9210120332, NRC Papers.

36. FWPCA Report, 4–6; *Colorado River Basin Hearing*, 10.

37. FWPCA Report, 7.

38. Ibid., 7–8; *Colorado River Basin Hearing*, 5–10.

39. Harold L. Price to W. B. McCool, "Proposed Discussion Paper: 'Ultimate Disposition of Uranium Mill Tailings,'" November 26, 1965, Accession No. 9210120324, NRC Papers; Glenn Seaborg to Edwin O. Wicks, and AEC *Staff Comments on Dr. Wicks' Letter*, June 29, 1966, Box 184, AEC Seaborg Papers; C. L. Henderson, memorandum to files, March 22, 1966, Accession No. 92101203360, NRC Papers.

40. Hollingsworth and Price to Seaborg et al., December 2, 1965.

41. Nussbaumer to files, January 7, 1966, Accession No. 9210120343, NRC Papers; Harold L. Price to John T. Conway, February 11, 1966, Box 660, JCAE Papers; John T. Conway, memorandum to "All [JCAE] Committee Members," February 18, 1966, Box 235, Aspinall Papers.

42. Nussbaumer to files, January 7, 1966; "West Slope Studying Pollution by Uranium," *Rocky Mountain News*, December 14, 1965.

43. A. O. Little, memorandum to Glenn Seaborg, March 2, 1966, Box 184, AEC Seaborg Papers.

44. FWPCA Report, 4; Conway to Aspinall, March 24, 1966, Box 235, Aspinall Papers and Box 169, JCAE Papers; *Colorado River Basin Hearing*, 17–20, 99; Rafford L. Faulkner, memorandum to George F. Quinn, April 15, 1966, Accession No. 9210120387, NRC Papers; Philippe G. Jacques to Gilbert A. Harrison, March 17, 1966, Box 268, Aspinall Papers.

45. Price to McCool, "Proposed Discussion Paper," November 26, 1965. Because the FWPCA and the USPHS worked together closely and were both under the Department of Health, Education and Welfare, the FWPCA was often referred to loosely as the "PHS."
46. FWPCA Report, 5–7; *Colorado River Basin Hearing*, 117–18; Report of the Ad Hoc Committee of the National Committee on Radiation Protection and Measurements, "Somatic Radiation Dose for the General Population," *Science* 131 (February, 1960): 482. The FWPCA report cites NCRP, National Bureau of Standards, *Maximum Permissible Body Burdens and Maximum Permissible Concentrations of Radionuclides in Air and Water for Occupational Exposure (Handbook 69)* (Washington, D.C.: Government Printing Office, June 1959); and Federal Radiation Council, *Radiation Protection Guidance for Federal Agencies*, Memorandum for the President, May 13, 1960.
47. "Uranium Mystery in the Colorado Basin," *New Republic* 154 (March 5, 1966): 9; *New Republic* 154 (April 16, 1966): 36–37. It is unclear how the *New Republic* was able to report on the contents of the FWPCA report in its March 5 edition when the report was not officially issued until almost three weeks later. One explanation is that someone leaked it to the magazine. It is also possible that the *New Republic* was able to piece together the article from available information about the contents of the report that were aired at the Grand Junction conference in December 1966.
48. Nussbaumer to the files, March 22, 1966, Accession No. 9210120360, NRC Papers; Harold L. Price, memorandum to Glenn Seaborg et al., March 4, 1966, Box 7717, AEC Secretariat Papers; Price to Conway, March 22, 1966, Box 235, Aspinall Papers; Jacques to Harrison, March 17, 1966.
49. Conway to Aspinall, March 13, 1967, Box 268, Aspinall Papers.
50. J. A. McBride, memorandum to Harold L. Price, April 13, 1966, Accession No. 9210120370, NRC Papers; Faulkner to Quinn, April 15, 1966; James B. Graham, memorandum to John T. Conway, April 14, 1966, Box 660, JCAE Papers; Conway to Aspinall, April 15, 1966, Box 169, JCAE Papers; George F. Quinn, memorandum to Glenn Seaborg et al., April 13, 1966, Box 184, AEC Seaborg Papers.
51. *Colorado River Basin Hearing*.
52. Ibid., 1–11. The one stabilized pile was at Monticello, Utah, at the site of the only federally owned uranium mill. When the mill closed in 1960, the tailings were graded, countered to enhance drainage, covered with rock and soil, and seeded. That stabilization program solved the problem of the tailings migrating onto adjoining property. Among private uranium millers, some actions were being explored to manage the tailings. One company, concerned about its pile located within a city, developed plans to cover its pile in a fashion similar to the Monticello pile. Another mill was stabilizing the sides of its tailings dam with rocks, soil, and vegetation. Most of the remaining piles near inhabited areas were kept moist to reduce wind erosion. Ibid., 4, 19, 41.
53. Ibid., 9–10.
54. Ibid., 20.
55. Ibid.
56. Conway to Aspinall, July 19, 1966, Box 169, JCAE Papers.
57. Frank C. Di Luzio to Glenn Seaborg, September 14, 1966, Box 184, AEC Seaborg Papers.
58. *Denver Post*, May 8, 1966, May 10, 1966; *Rocky Mountain News*, April 29, 1966, May 7, 1966; Grand Junction, Colo., *Daily Sentinel*, March 27, 1966; Sheldon Novick, "Radioactive Mining Wastes," *Scientist and Citizen* 8 (August 1966): 10–12; "Joint Federal Agency Position Regarding Control of Uranium Mill Tailings," December 8, 1966, Box 281, Aspinall Papers; Conway to Aspinall, December 20, 1966, Box 235, Aspinall Papers.

59. Philip R. Lee to Frank C. Di Luzio, November 25, 1966 (AEC 544/33, December 21, 1966), Box 7717, AEC Secretariat Papers.
60. Rafford L. Faulkner, memorandum to John A. Erlewine, December 13, 1966 (AEC 544/33, December 21, 1966), Box 7717, AEC Secretariat Papers; Conway to Aspinall, February 10, 1967, Box 281, Aspinall Papers.
61. Conway to Aspinall, February 10, 1967.
62. Roy L. Cleere to Wayne Aspinall, May 2, 1966, Box 235, Aspinall Papers, and Box 660, JCAE Papers; *Colorado River Basin Hearing*, 129–30.
63. *Colorado River Basin Hearing*, 98.
64. Faulkner to Quinn, June 7, 1966 (AEC 544/22, June 9, 1966), Box 7717, AEC Secretariat Papers. Mills operating in Wyoming and New Mexico were not contacted because their piles were considered to be too far away from established communities. They were to be contacted at a later date once an effective tailings management policy was established. A. T. F. Seal to Faulkner, July 28, 1966 (AEC 544/24, September 29, 1966), Box 7717, AEC Secretariat Papers; James B. Graham, memorandum to the files, May 9, 1966, Box 660, JCAE Papers.
65. *Colorado River Basin Hearing*, 113; Roy L. Cleere and William F. McGlone to Byron W. Rogers, March 16, 1967, Box 7717, AEC Secretariat Papers; William F. McGlone and Roy L. Cleere to Edmund S. Muskie, February 15, 1967, Box 7717, AEC Secretariat Papers; Colorado Department of Public Health, *Uranium Wastes*, 16–17.
66. *Regulations of the Colorado State Department of Public Health Requiring Stabilization of Uranium and Thorium Mill Tailing Piles*, May 9, 1966, and December 12, 1966, Box 9934, Colorado State Archives (*Radiation Regulation No. 2*); Conway to Aspinall, February 10, 1967.
67. J. A. McBride, memorandum to Harold L. Price, December 1, 1966, Accession No. 9210120402, NRC Papers. In the final version of the regulation, that language was later changed to "Access to the stabilized pile area shall be controlled by the operator or owner and properly posted."
68. Graham to Conway, May 17, 1966, Box 660, JCAE Papers; Press Release, Colorado Department of Public Health, April 14, 1966, Colorado State Historical Society, Denver; *Regulations of the Colorado State Department of Public Health Requiring Stabilization of Uranium and Thorium Mill Tailing Piles*; Cleere to Aspinall, December 22, 1966, Box 169, JCAE Papers; Roy L. Cleere to Glenn Seaborg, September 21, 1966 (AEC 544/29, October 21, 1966), Box 7717, AEC Secretariat Papers; McGlone and Cleere to Muskie, February 15, 1967. The AEC noted that the proposed Colorado regulations required control over tailings based on their radium and other radionuclide content, not because of their uranium or thorium content. Furthermore, the level of radium necessary to trigger state control was low. The AEC speculated that the requirement demonstrated the intent of the Colorado Public Health Department to control radioactive tailings from all sources. The AEC highlighted two gold tailings piles that contained radium in excess of the Colorado threshold and would, therefore, be subject to regulation. Donald Nussbaumer, memorandum to Wayne Kerr, November 10, 1966, Accession No. 9210120394, NRC Papers.
69. Conway to Aspinall, February 10, 1967.

Chapter 5

1. Glenn Seaborg to Edwin O. Wicks, June 29, 1966, Box 184, U.S. Atomic Energy Commission Glenn Seaborg Collection, U.S. Department of Energy Archives, Germantown, Md. (hereafter AEC Seaborg Papers).

2. William F. McGlone and Roy L. Cleere to Edmund Muskie, February 15, 1967, Box 7717, U.S. Atomic Energy Commission Secretariat Collection, U.S. Department of Energy Archives, Germantown, Md. (hereafter AEC Secretariat Papers); William F. McGlone and Roy L. Cleere to Byron W. Rogers, March 16, 1967, Box 7717, AEC Secretariat Papers.
3. John A. Erlewine, memorandum to Robert E. Hollingsworth and Harold L. Price, November 8, 1967 (AEC 544/75, November 14, 1967), Box 7717, AEC Secretariat Papers.
4. "Joint Federal Agency Position Regarding Control of Uranium Mill Tailings," December 8, 1966, Box 281, Wayne Aspinall Collection, University of Denver Penrose Library, Denver, Colo. (hereafter Aspinall Papers); John T. Conway, memorandum to Wayne Aspinall, December 20, 1966, Box 235, Aspinall Papers; Conway to Aspinall, July 19, 1966, Box 169, Joint Committee on Atomic Energy General Correspondence File, Record Group 128, National Archives and Records Administration, Washington, D.C. (hereafter JCAE Papers); Conway to Aspinall, February 10, 1967, Box 281, Aspinall Papers.
5. "AEC Radiation Control Program for Uranium Mill Operations," May 11, 1960, 1–5, Box 251, JCAE Papers; Philip R. Lee to Frank C. Di Luzio, November 25, 1966 (AEC 544/33, December 21, 1966), Box 7717, AEC Secretariat Papers; Martin B. Biles, memorandum to R. E. Hollingsworth, December 13, 1966, Box 7717, AEC Secretariat Papers.
6. Sigismund Peller, "Lung Cancer among Mine Workers in Joachimsthal," *Human Biology* 11 (1939): 130–43; W. C. Hueper, *Occupational and Environmental Cancer of the Respiratory System* (Springfield, Ill.: Charles C. Thomas, 1942), 435–56; Egon Lorenz, "Radioactivity and Lung Cancer: A Critical Review of Lung Cancer in the Miners of Schneeberg and Joachimsthal," *Journal of the National Cancer Institute* 5 (August 1944): 1–15; E. Cook, "Ionizing Radiation," in W. W. Murdock, ed., *Environment* (Sunderland, Mass.: Sinauer Associates, 1975), 304; "Chronology of Uranium Mining Health Protection Activities," November 1970 (SECY-934, January 22, 1971), Box 7822, AEC Secretariat Papers.
7. Rhonda S. Berger, "The Carcinogenicity of Radon," *Environmental Science and Technology* 24, no. 1 (1990): 30–31; National Research Council, Committee on the Biological Effects of Ionizing Radiations, *Health Risks of Radon and Other Internally Deposited Alpha-Emitters, Beir IV* (1988). Because radon 219 (actinon) and radon 220 (thoron) have very short half-lives, they exist in low concentrations and pose less danger. Alpha particles have high mass but low penetrating power. Inside the lungs, however, they can cause considerable damage for the short distance they travel.
8. Betty L. Perkins, *An Overview of the New Mexico Uranium Industry* (Santa Fe: New Mexico Energy and Minerals Department, January 1979), 98, 129; Ken Silver, "The Yellowed Archives of Yellowcake," *Public Health Reports* 111 (March–April 1996): 116–27; Joseph Wagoner, "Uranium Mining and Milling: The Human Costs," text of a speech given at the University of New Mexico Medical School, Albuquerque, March 10, 1980; and Douglas G. Peter, M.D. (U.S. Department of Health, Education and Welfare, Public Health Service), Memorandum to "Whom it May Concern," June 4, 1979 (both documents are located at the Southwest Research and Information Center, Albuquerque, N.Mex.); Richard Waxweiler et al., "Mortality Patterns among a Retrospective Cohort of Uranium Mill Workers," *Proceedings of the Sixteenth Midyear Topical Meeting of the Health Physics Society*, Albuquerque, January 9–13, 1983.
9. Raye C. Ringholz, *Uranium Frenzy: Boom and Bust on the Colorado Plateau* (New York: Norton, 1989), 41–43. Ringholz quotes from Batie's sworn testimony in *John N.*

Begay, et al. v. The United States of America, 591 F. Supp. 991 (1984), correspondence, and a personal interview.
10. Ringholz, *Uranium Frenzy,* 41–43.
11. Ibid., 41–45; H. Peter Metzger, *The Atomic Establishment* (New York: Simon and Schuster, 1972), 164 (quoting P. W. Jacoe, December 1970); Silver, "The Yellowed Archives of Yellowcake," 124.
12. Ringholz, *Uranium Frenzy,* 44.
13. Rafford L. Faulkner, memorandum to Glenn Seaborg et al., October 20, 1965, Box 7717, AEC Secretariat Papers. "Note by the Secretary," October 11, 1965, Accession No. 9210120270, U.S. Nuclear Regulatory Commission Public Documents Reading Room, Washington, D.C., 4, 9, 10–11 (hereafter NRC Papers).
14. Harold L. Price to John T. Conway, February 11, 1966, Box 235, Aspinall Papers; Rafford L. Faulkner, memorandum to Glenn Seaborg et al., October 20, 1965.
15. Colorado Department of Public Health, *Uranium Wastes,* 19, citing G. F. Tape to Colorado Department of Public Health, June 9, 1966; Price to Conway, February 11, 1966; Faulkner to Seaborg et al., October 20, 1965; L. K. Olson, memorandum to Harold L. Price, April 15, 1960, Accession No. 9210120199, NRC Papers.
16. "Wigton Lists VCA Pile Health Hazard," *Durango Herald,* May 10, 1966.
17. Chester M. Wigton et al. to Roy L. Cleere, May 15, 1967, Box 268, Aspinall Papers.
18. Faulkner to Seaborg et al., October 20, 1965; Price to Conway, February 11, 1966; U.S. Congress, Subcommittee on Air and Water Pollution of the Senate Committee on Public Works, *Radioactive Water Pollution in the Colorado River Basin,* 89th Cong., 2d sess., May 6, 1966, 41, 45 (hereafter *Colorado River Basin Hearing*).
19. Lee to Di Luzio, November 25, 1966.
20. Biles to Hollingsworth, December 13, 1966.
21. Colorado Department of Public Health, *Uranium Wastes,* 15; Colorado Department of Public Health, "Report on Control of Uranium Mill Tailings," October 28, 1966, Box 9937, Colorado Archives, 5; Biles to Hollingsworth, December 13, 1966; J. A. McBride, memorandum to Harold L. Price, December 1, 1966, Accession No. 9210120402, NRC Papers; Glenn Seaborg to Wayne Aspinall, February 10, 1967, Box 281, Aspinall Papers (see "Staff Report on Tailings Status" [undated], attached to letter); Robley D. Evens to J. L. Robison, July 26, 1966 (AEC 544/24, September 9, 1966), Box 7717, AEC Secretariat Papers.
22. Colorado Department of Public Health, *Uranium Wastes,* 16; Martin B. Biles memorandum to E. J. Bloch, March 7, 1967 (AEC 544/40, March 13, 1967), Box 7717, AEC Secretariat Papers; Vernon G. MacKenzie and Martin B. Biles, "Evaluation of Radon from Uranium Mill Tailings," July 1967 (AEC 544/64, August 2, 1967), Box 7717, AEC Secretariat Papers; Robert E. Hollingsworth to Wayne Aspinall, May 22, 1967, Box 661, JCAE Papers.
23. Hollingsworth to Aspinall, May 22, 1967; Conway to Aspinall, February 10, 1967; J. Samuel Walker, *Containing the Atom: Nuclear Regulation in a Changing Environment, 1963–1971* (Berkeley: University of California Press, 1992), 261.
24. Wayne Aspinall to Glenn Seaborg and John Gardner, February 9, 1967, Box 281, Aspinall Papers.
25. Seaborg to Aspinall, February 28, 1967, Box 661, JCAE Papers; Seaborg to Aspinall, February 28, 1967 (AEC 544/40, March 13, 1967), Box 7717, AEC Secretariat Papers.
26. *Regulations of the Colorado State Department of Public Health Requiring Stabilization of Uranium and Thorium Mill Tailing Piles, Radiation Regulation,* May 9, December 12, 1966, Box 9934, Colorado State Archives.

27. Colorado Department of Public Health Press Release, June 1, 1967, State Department of Health File, 1967–68, Colorado Historical Society, Denver, Colo.; Colorado Department of Public Health, *Uranium Wastes*, 21. Durango residents were indignant that they would have to wait until 1970 to see the VCA pile stabilized. Wigton to Cleere, May 15, 1967; John A. Erlewine, memorandum to Robert E. Hollingsworth, November 8, 1967, Box 7717, AEC Secretariat Papers.
28. Roy L. Cleere to Glenn Seaborg, April 25, 1966, Box 660, JCAE Papers, and Box 235, Aspinall Papers.
29. Erlewine to Hollingsworth, November 8, 1967.
30. Ibid.
31. Colorado Department of Public Health, *Uranium Wastes*, 19, citing G. F. Tape to Colorado Department of Public Health, June 9, 1966; "AEC Staff Comments on Dr. Wicks' Letter," June 29, 1966, Box 184, AEC Seaborg Papers; Biles to Hollingsworth, August 4, 1967 (AEC 544/65, August 9, 1966), Box 7717, AEC Secretariat Papers.
32. Rafford L. Faulkner, memorandum to George F. Quinn, June 7, 1966 (AEC 544/22, June 9, 1966), Box 7717, AEC Secretariat Papers; Faulkner to Quinn, August 30, 1966 (AEC 544/24, September 9, 1966), Box 7717, AEC Secretariat Papers; Glenn Seaborg to Edwin O. Wicks, June 29, 1966, Box 184, AEC Seaborg Papers; John T. Conway memorandum to JCAE, February 18, 1966, Box 235, Aspinall Papers; Price to Conway, February 11, 1966, Box 660, JCAE Papers.
33. "Statement of the Atomic Energy Commission, Conference on Pollution of the Colorado River, Sixth Session" (Denver, Colo.), July 26, 1967 (AEC 544/65, August 9, 1967), Box 7717, AEC Secretariat Papers.
34. Erlewine to Hollingsworth, November 8, 1967.
35. Robert E. Hollingsworth and Harold L. Price to Glenn Seaborg et al., December 20, 1968 (AEC 544/90, December 23, 1968), Box 7764, AEC Secretariat Papers (see attachment to letter: "Atomic Energy Commission Status Report, Uranium Mill Tailings," [undated]), 8–9, 31–32.
36. Ibid., 9.
37. Colorado Department of Public Health, *Uranium Wastes*, 23, citing J. M. Rademacher, "Memorandum to Interagency Technical Committee in Control of Uranium Mill Tailings Piles," February 7, 1968; Biles to Hollingsworth, December 13, 1966.
38. In 1961 and 1962 AEC stabilized its own Monticello pile with rock, soil, vegetation, and various contouring measures to minimize erosion.
39. Hollingsworth and Price to Glenn Seaborg et al., December 20, 1968, 6–8.
40. U.S. Department of Health, Education, and Welfare, U.S. Public Health Service, *Evaluation of Radon 222 Near Uranium Tailings Piles*, DER 69–1 (Rockville, Md.: U.S. Department of Health, Education and Welfare, March 1969); Roy L. Cleere to Wayne Aspinall, October 8, 1969 (AEC 544/106, November 14, 1969), Box 7764, AEC Secretariat Papers.
41. Hollingsworth and Price to Seaborg et al., December 20, 1968, 6–8.
42. U.S. Public Health Service, *Evaluation of Radon 222*, 20; Biles to Hollingsworth, June 6, 1969 (AEC 544/100, June 11, 1969), Box 7764, AEC Secretariat Papers; Cleere to Aspinall, October 8, 1969.
43. Hollingsworth and Price to Seaborg et al., December 20, 1968.

Chapter 6

1. Samuel P. Hays, *Beauty, Health, and Permanence: Environmental Politics in the United States, 1955–1985* (Cambridge: Cambridge University Press, 1987), 54; Allen V.

Kneese and Charles L. Schultze, *Pollution, Prices, and Public Policy* (Washington, D.C.: Brookings Institution Press, 1975), 2–3.
2. U.S. Environmental Protection Agency, Office of Radiation Programs, *Outdoor Radon Study (1974–1975): An Evaluation of Ambient Radon-222 Concentrations in Grand Junction, Colorado*, Technical Note ORP/LV-77-1 (April 1977).
3. U.S. Congress, Senate, Committee on Public Works, Subcommittee on Air and Water Pollution, *Radioactive Water Pollution in the Colorado River Basin*, 89th Cong., 2d sess., May 6, 1966, 20 (hereafter *Colorado River Basin Hearing*).
4. Other cities suffering from significant pollution resulting from tailings used in construction include Durango, Colorado, Salt Lake City, Utah, and Shiprock and Tuba City, New Mexico. Not all of the affected cities were in the West, however. Canonsburg, Pennsylvania, had been the site of the Vitro Rare Metals Company, which had begun processing uranium ore for radium in 1911 and later processed ore for uranium oxide for the AEC. Ultimately, uranium processing at this site generated 200,000 tons of tailings. In 1966 the AEC informed Vitro that "due to the insignificance of the contamination which may be present in your former Canonsburg facility in this particular instance, no hazard to health and safety is involved as a result of AEC licensed activities." Donald A. Nussbaumer to Vitro Corporation of America, February 14, 1966, located in Congress, House, Subcommittee on Energy and the Environment of the Committee on Interior and Insular Affairs, *Uranium Mill Tailings Control: H.R. 13382, H.R. 12938, H.R. 12535 and H.R. 13049*, 95th Cong., 2d sess., June 26–27, July 10, 17, 1978, Appendix 1, 179 (hereafter *Uranium Mill Tailings Control Hearing*). In 1967 the tailings site was developed into an industrial park, and thirteen years later the buildings on the eighteen-acre location were razed and the radioactive tailings were removed. Ultimately, one hundred Canonsburg structures, including a swimming pool and a skating rink, were found to be so contaminated that they were destroyed and buried. Ellen Wilson, "Some Like it Hot," *Environmental Action* 17 (November–December 1985): 28–32; "Nuclear Waste Kills Investment," *Business Week* (March 17, 1980): 41; Ralph Haurwitz, "Families Cry for Radiation Park Action," *Pittsburgh Press*, October 26, 1980. In the Palos Park area near Chicago, the AEC examined 150 structures that had been turned over to the private sector. The Commission discovered that thirty of those buildings had been contaminated by tailings from the Middlesex tailings site and required cleanup. *Uranium Mill Tailings Control Hearing*, 42. The problem of tailings used for construction continues to reappear. A home in Whiteford Township, Michigan, underwent a $2 million cleanup in 1994 when it was discovered that it had been landscaped with soil brought from the site of a Manhattan Project facility in Toledo, Ohio. *Ann Arbor News*, October 26, 1994.
5. "Landscaping (Industrial Strength)," *Nuclear Energy* (Second Quarter 1993): 9–11; U.S. Federal Power Commission, *National Power Survey, Environmental Research (The Report and Recommendations of the Task Force on Environmental Research to the Technical Advisory Committee on Research and Development)* (Washington, D.C.: Government Printing Office, January, 1974), VI-65; J. M. Costello et al., "A Review of the Environmental Impact of Mining and Milling of Radioactive Ores, Upgrading Processes, and Fabrication of Nuclear Fuels," in Essam E. El-Hinnawi, ed., *Nuclear Energy and the Environment* (Oxford: Pergamon Press, 1980), 45–47; "Management of Inactive Uranium Mill Tailings," *Journal of Environmental Engineering* 112 (June 1986): 507–9; R. M. Fry, "Criteria for the Long-Term Management of Uranium Mill Tailings," in *Management of Wastes from Uranium Mining and Milling: Proceedings of an*

International Symposium on the Management of Wastes from Uranium Mining and Milling (Vienna: International Atomic Energy Agency, 1982), 81–82; D. Lush et al., "An Assessment of the Long Term Interaction of Uranium Tailings with the Natural Environment," *Proceedings of the Seminar on Management, Stabilization and Environmental Impact of Uranium Mill Tailings* (Albuquerque, N.Mex.: OECD Nuclear Energy Agency, July 1978). One tailings pile stabilized in 1963 was eroded by gophers. Roger Rapoport, "The Trouble with 90.5 Million Tons of Radioactive Tailings," *Los Angeles Times (West Magazine)*, April 12, 1970, 10.

6. U.S. Department of Health, Education and Welfare, *Disposition and Control of Uranium Mill Tailings Piles in the Colorado River Basin* (Denver: Federal Water Pollution Control Administration, Region VII, March 1966), 8, Box 235, Wayne Aspinall Papers, Penrose Library, University of Denver, Denver, Colo. (hereafter Aspinall Papers).

7. Earl F. Peterson to Wallace Bennett, April 14, 1965, Box 171, Joint Committee on Atomic Energy General Correspondence File, Record Group 128, National Archives and Records Administration, Washington, D.C. (hereafter JCAE Papers).

8. U.S. Congress, Joint Committee on Atomic Energy, Subcommittee on Raw Materials, *Use of Uranium Mill Tailings for Construction Purposes*, 92d Cong., 1st sess., October 28–29, 1971, 143 (hereafter *Tailings for Construction Hearing*).

9. Ibid., 146–47; Martin B. Biles, memorandum to files, October 2, 1970 (SECY-934, January 22, 1971), Box 7822, U.S. Atomic Energy Commission Secretariat Collection, U.S. Department of Energy Archives, Germantown, Md. (hereafter AEC Secretariat Papers); *Fruita* (Colo.) *Times*, March 18, 1982.

10. Conservative estimates ranged from 150,000 to 200,000 tons. Colorado Department of Public Health, *Uranium Wastes and Colorado's Environment*, 2d ed. (Denver: Colorado Department of Public Health, August 1971), 29; *Tailings for Construction Hearing*, 106, 145, 147. The 300,000-ton estimate from the AMAX Corp. (successor to the Climax corporation) is probably the most accurate estimate because it was made by longtime mill managers, one of whom had worked at the mill since 1952. The company, moreover, had little incentive to downplay the statistics or favor any of the parties involved in evaluating the tailings problem in Grand Junction. At the time, AMAX was not considered to have been liable for this problem and the company was cooperating fully with the health investigations.

11. Colorado Department of Public Health, "Report on Control of Uranium Mill Tailings," October 28, 1966, Box 9937, Colorado State Archives, Denver, 5; Robert N. Snelling and Robert D. Siek, "Evaluation of Radon Film Badge" (Southwestern Radiological Health Laboratories, USPHS, and Colorado Department of Public Health, April 1968); Colorado Department of Public Health, *Uranium Wastes*, 25; *Tailings for Construction Hearing*, 42–43, 100, 107–8; H. Peter Metzger, *The Atomic Establishment* (New York: Simon and Schuster, 1972), 171.

12. R. J. Augustine, "Uranium Mill Tailings Used for Construction Fill: A Summary of the Problem and Current Bureau of Radiological Health Activities," October 22, 1970, Box 661, JCAE Papers; Edward J. Bauser, memorandum to the files, August 10, 1970, Box 661, JCAE Papers; Roy L. Cleere to Wayne Aspinall, October 8, 1969 (AEC 544/106, November 14, 1969), Box 7764, AEC Secretariat Papers; "Uranium Legacy," *The Workbook* 8 (November–December 1983): 192–95. Investigators used direct gamma radiation measurements to locate tailings deposits in residential areas, as well as natural outcroppings of other radioactive materials, in a reasonably short time and with reasonable accuracy.

13. R. J. Augustine, "Uranium Mill Tailings Used for Construction Fill"; Bauser to the files, August 10, 1970; Cleere to Aspinall, October 8, 1969; H. W. Lewis, *Technological Risk* (New York: Norton, 1990), 247–53; "Uranium Legacy," *The Workbook* 8 (November–December 1983): 192–95; *Tailings for Construction Hearing*, 47, 71; Grand Junction *Daily Sentinel*, March 25, 1971; U.S. Congress, Joint Committee on Atomic Energy, Special Subcommittee on Radiation, *Hearings on Radiation Protection Criteria and Standards: Their Basis and Use*, 86th Cong., 2d sess., May 24–26, 31, June 1–3, 1960, 187, 220.
14. Anthony Mastrovich to Paul Jacoe, January 27, 1966, Box 426, Aspinall Papers; "Statement of Anthony Mastrovich, October 29, 1971," Box 426, Aspinall Papers; Paul Jacoe to Anthony Mastrovich, February 7, 1966, Box 426, Aspinall Papers.
15. *Tailings for Construction Hearing*, 42–43, 100, 148. By 1971 all states except Wyoming (which lacked legal authority) and Oregon (which had regulations pending) had forbidden the use of tailings in construction. Bauser to the files, August 10, 1970; Frank E. McGinley, memorandum to Elton A. Youngberg, November 16, 1971, Accession No. 9210120418, U.S. Nuclear Regulatory Commission Public Documents Reading Room, Washington, D.C. (hereafter NRC Papers).
16. Colorado Department of Public Health, *Uranium Wastes*, 25.
17. Ibid., 1, 16, 22, 25, 34; U.S. Public Health Service, Department of Health, Education and Welfare, *Evaluation of Radon 222 Near Uranium Tailings Piles*, DER 69-1 (March 1969); Martin B. Biles, memorandum to John A. Erlewine, August 25, 1970 (SECY-332, September 8, 1970), Box 7822, AEC Secretariat Papers; W. E. Johnson to Thomas Eagleton, March 30, 1970, Box 7764, AEC Secretariat Papers; H. Peter Metzger, "'Dear Sir: Your House Is Built on Radioactive Uranium Waste,'" *New York Times Magazine*, October 31, 1971, 63; Metzger, *Atomic Establishment*, 180.

    Radon measurements are expressed in picocuries per liter. Radon daughter measurements are expressed in "working level," or WL. These units represent the amount of radon or daughter atoms in a liter of air or, equivalently, as activities—that is, a measure of the rate at which radon and its daughters are decaying in the air sample. A WL is defined as any combination of radon daughters in 1 liter of air that will result in the ultimate emission of 1.3×10^5 MeV of potential alpha energy. This numerical value is derived from the alpha energy released by the total decay of the short-lived radon daughter products at equilibrium with 100 picocuries of radon 222 per liter. One picocurie per liter is a concentration of 18,000 radon atoms in the sample, or an activity of 2.2 radon atoms decaying per minute in the sample. The WL measurement of radon daughters offers a rough measure of radon hazards. In houses, 100 picocuries per liter is approximately 1 WL. Michael R. Edelstein and William J. Makofske, *Radon's Deadly Daughters: Science, Environmental Policy and the Politics of Risk* (Lanham, Md.: Rowman and Littlefield, 1998), 5.
18. U.S. Atomic Energy Commission, "Indoor Radon Daughters and Radiation Measurements in East Tennessee and Central Florida," HASL-TM-71-8 (New York: Health and Safety Laboratory, AEC), March 1971; *Tailings for Construction Hearing*, 65, 104; Metzger, "'Dear Sir'" 63; Metzger, *Atomic Establishment*, 180.
19. Stephen H. Greenleigh, memorandum to Legal Files, December 18, 1970, NRC Accession No. 9210120373; *Draft: U.S. Atomic Energy Commission Statement for JCAE Hearings on the Use of Uranium Mill Tailings*, October 28–29, 1971, Box 426, Aspinall Papers.
20. Robert E. Hollingsworth and Harold L. Price, memorandum to Glenn Seaborg, December 20, 1968 (AEC 544/90, December 23, 1968), Box 7764, AEC Secretariat

Papers; Colorado Department of Public Health, *Uranium Wastes,* 29–30; *Tailings for Construction Hearing,* 71–72; Biles to Erlewine, June 17, 1970 (SECY-191, August 5, 1970), Box 7822, AEC Secretariat Papers; Anthony Ripley, "City in Colorado Awakens to Scope of Radioactive Waste Problem," *New York Times,* October 4, 1971; *Newsweek* 78 (October 18, 1971): 46; *Time* 98 (December 20, 1971): 56.

21. *Rocky Mountain News,* September 1, 1966; *Washington Daily News,* September 1, 1966.
22. J. Samuel Walker, *Containing the Atom: Nuclear Regulation in a Changing Environment, 1963–1971* (Berkeley: University of California Press, 1992), 262; *Tailings for Construction Hearing,* 113, 135, 242–43; Cleere to Aspinall, October 8, 1969; Donald A. Nussbaumer, memorandum to the files, March 18, 1964, Accession No. 9210120332, NRC Papers; Grand Junction *Daily Sentinel,* January 13, March 25, 26, 29, 1970; Rafford L. Faulkner, memorandum to Glenn Seaborg, April 2, 1970 (AEC 544/109, April 7, 1970), Box 7764, AEC Secretariat Papers; *Denver Post,* December 18, 19, 1969; *Rocky Mountain News,* December 20, 1969; Rapoport, "The Trouble with 90.5 Million Tons of Radioactive Tailings."
23. N. Wood, "America's Most Radioactive City," *McCall's* 97 (September 1970): 46.
24. "Radon? Sure. So What Else Is New? Ask the Folks at Grand Junction," *Nuclear Industry* 18, nos. 10–11, pt. 2 (October–November 1971): 34–39; "Progress Report, Colorado Indoor Radon Study" (undated), (SECY-1806, June 29, 1971), Box 7822, AEC Secretariat Papers.
25. *Tailings for Construction Hearing,* 134–42; Bauser to the files, August 10, 1970; Progress Report, Colorado Indoor Radon Study" (undated) (SECY-1806, June 29, 1971), Box 7822, AEC Secretariat Papers; John A. Erlewine, memorandum to James Schlesinger, August 20, 1971, Box 7822, AEC Secretariat Papers; "Radon? Sure. So What Else Is New?"; Grand Junction *Daily Sentinel,* March 25, 26, 29, 1970, September 9, 1971; Steve Wynkoop, "Panel Asks Removal of Junction Tailings," *Denver Post,* September 22, 1971; "Homeowners Show Little Concern," *Denver Post,* November 25, 1971; Ripley, "City in Colorado Awakens to Scope of Radioactive Waste Problem"; *Newsweek* 78 (October 18, 1971): 46; *Time* 98 (December 20, 1971): 56.
26. Glenn Seaborg to John A. Love, March 31, 1971, in W. B. McCool, "Proposed University of Colorado Contract On Effects of Radioactive Tailings," April 9, 1971 (SECY-1325), Box 7822, AEC Secretariat Papers; Anthony Ripley, "Infants and Radioactive Sands: Small-Town Doctor Wins Fight," *New York Times,* October 3, 1971; Anthony Ripley, "Did Someone Make a Deadly Blunder?" *New York Times (Week in Review),* October 3, 1971.
27. *Tailings for Construction Hearing,* 281–99; "Hot Town," *Time* (December 20, 1971): 56; Martin B. Biles, memorandum to Robert E. Hollingsworth, September 8, 1971, Box 7822, AEC Secretariat Papers.
28. *Tailings for Construction Hearing,* 308–18; S. G. English to Donald G. Brotzman, November 5, 1971, Box 7822, AEC Secretariat Papers. A later health study conducted by the Colorado Department of Public Health for the NRC found that there was no noticeably higher incidence of lung cancer in Mesa County and that the slight but statistically insignificant increase in lung cancer among males aged 35 to 49 was attributed to the presence of uranium miners in the study population. The study did show a twofold excess incidence of leukemia in Mesa County, distributed equally among males and females. Persons aged 65 and over were affected by the disease two and a half times the expected rate. The leukemia was of a type (acute myelogenous) that was often seen in excess in populations exposed to radiation. Although the researchers did

not directly associate the cancer with the tailings, they noted that the cancer rate raised "serious questions in a geographical area where radioactive uranium mill tailings have been used extensively for construction purposes." M. C. Cunningham, S. W. Ferguson, and T. Foreman, *Excess Cancer Incidence in Mesa County, Colorado* (n.d.), prepared for the U.S. Nuclear Regulatory Commission by the Colorado Department of Public Health, 17–18.

29. Cleere to Aspinall, October 8, 1969; *Tailings for Construction Hearing*, 7; J. A. McBride, memorandum to Harold L. Price, March 27, 1969, Accession No. 9210120405, NRC Papers.
30. Biles to Erlewine, June 17, 1970, Accession No. 9210120413, NRC Papers.
31. *Tailings for Construction Hearing*, 7, 55, 397–98. McBride to Price, March 11, 1969, Accession No. 9210120378, NRC Papers; Cleere to Aspinall, October 8, 1969; Robert Saile, "Radon Gas Found in Junction Homes," *Denver Post*, December 19, 1969; Biles to Erlewine, June 17, 1970; George D. Aiken to Glenn Seaborg, January 30, 1970 (AEC 544/107, February 25, 1970), Box 7764, AEC Secretariat Papers; Aiken to Seaborg, January 30, 1970; E. J. Bloch to Aspinall, November 14, 1969 (AEC 544/106, November 14, 1969), Box 7764, AEC Secretariat Papers; Rafford. L. Faulkner, memorandum to Robert E. Hollingsworth, September 24, 1970, Accession No. 9210120249, NRC Papers.
32. "Progress Report, Colorado Indoor Radon Study" (undated), (SECY-934, January 22, 1971), Box 7822, AEC Secretariat Papers; "Colorado Indoor Radon Study, Second Meeting of Joint Agency Working Committee," December 7–8, 1970 (SECY-934, January 22, 1971), Box 7822, AEC Secretariat Papers; "Colorado Indoor Radon Study: First Meeting of Joint Agency Working Committee," December 31, 1970, Accession No. 9210120237, NRC Papers; "Progress Report, Colorado Indoor Radon Study" (undated), (SECY-1806, June 29, 1971), Box 7822, AEC Secretariat Papers.
33. Roy L. Cleere to M. W. Carter, January 29, 1970 (SECY-332, September 8, 1970), Box 7822, AEC Secretariat Papers, and (AEC 544/107, February 25, 1970), Box 7764, AEC Secretariat Papers.
34. FRC guidelines for annual whole-body exposures to gamma radiation for individuals in the general population were 0.5 rem per year. (Rem, or roentgen equivalent man, is the measure of biological damage from radiation.) Continuous exposure to 0.05 milliroentgen per hour would result in an estimated whole-body dose of 0.5 rem per year. For radon daughters, the FRC and ICRP (revised) guidelines for occupational exposure were 4 WLM per year. (WLM refers to "working level month" and is a product of the average rate of exposure in working levels and the number of 170-hour working months of exposure.) Continuous exposure (165 hours per week) at the rate of 0.05 WL could result in an exposure of about one-half the recommended occupational exposure level. Martin B. Biles, letter to Robert E. Hollingsworth, February 20, 1970 (AEC 544/107, February 25, 1970), Box 7764, AEC Secretariat Papers.
35. *Tailings for Construction Hearing*, 77, 309; Colorado Department of Public Health, *Uranium Wastes*, 34–35; *Environmental Reporter* 1 (August 21, 1970): 446–47.
36. *Tailings for Construction Hearing*, 7, 52–54, 76, 159–60; Colorado Department of Public Health News Release, August 11, 1970, Box 184, Atomic Energy Commission Glenn Seaborg Collection, United States Department of Energy Archives, Germantown, Md. (hereafter AEC Seaborg Papers); Paul J. Peterson to Roy L. Cleere, July 27, 1970 (SECY-332, September 8, 1970), Box 7822, AEC Secretariat Papers; Bauser to the files, August 10, 1970; Martin B. Biles to Roy L. Cleere, September 4, 1970 (SECY-350, September 10, 1970), Box 7822, AEC Secretariat Papers; "AEC Staff Analysis, PHS Recommendations of July 27, 1970 on Action Levels Associated with

Uranium Tailings in Dwellings," August 26, 1970 (SECY-350, September 10, 1970), Box 7822, AEC Secretariat Papers; U.S. Congress, House, "Statement by Congressman Aspinall Concerning Use of Mill Tailings," *Congressional Record* 116, pt. 21 (August 12, 1970): 28579–28581.

37. *Tailings for Construction Hearing*, 8; R. J. Augustine to Edward J. Bauser, October 26, 1970, Box 661, JCAE Papers; "Colorado Indoor Radon Study: First Meeting of Joint Agency Working Committee," December 31, 1970, Accession No. 9210120237, NRC Papers.

38. Grand Junction *Daily Sentinel*, September 9, 1971.

39. Faulkner to Hollingsworth, September 24, 1970; James Schlesinger to George P. Shultz (undated) (SECY-2359, March 13, 1972), Box 7822, AEC Secretariat Papers; "Proposed Policy Issues for Commission Consideration, etc.," October 12, 1971 (SECY-2117, October 12, 1971), Box 7822, AEC Secretariat Papers, 4; Biles to Hollingsworth, September 8, 1971.

40. *Tailings for Construction Hearing*, 104–5; Faulkner to Hollingsworth, September 24, 1970; Faulkner, memorandum to the file, April 3, 1971, Accession No. 9210120367, NRC Papers; Glenn Seaborg, letter to George P. Schultz, July 21, 1971, Box 7822, AEC Secretariat Papers. "Progress Report, Colorado Indoor Radon Study" (undated) (SECY-1806, June 29, 1971), Box 7822, AEC Secretariat Papers. This report explored the possibility of having all the federal agencies with an interest in solving the indoor radiation problem (AEC, HEW, EPA, HUD) share the costs if federal funding was necessary.

41. Bauser to the files, August 10, 1970; *Colorado River Basin Hearing*; Wayne Aspinall and Peter Dominick to John O. Pastore, September 21, 1971, Box 7822, AEC Secretariat Papers; Press Release No. 663, "JCAE Subcommittee on Raw Materials Announces Hearings on Use of Uranium Mill Tailings," October 6, 1971, Box 226, JCAE Papers.

42. Tom Rees, "Committee Recommends Uranium Tailings Action," *Rocky Mountain News*, September 22, 1971.

43. E. J. Bloch to Edward J. Bauser (and attached "Synopsis of Deliberations"), September 30, 1971, Box 662, JCAE Papers; *Tailings for Construction Hearing*, 9–10, 78–80, 101–2, 120; Erlewine to Schlesinger, September 29, 1971; Wayne Aspinall to William D. Ruckelshaus, September 24, 1971, Box 7822, AEC Secretariat Papers; Alice Wright, "Tailings Removal Urged at Government's Expense," Grand Junction *Daily Sentinel*, September 21, 1971; Anthony Ripley, "Radioactive Building Sand Stirs Dispute," *New York Times*, September 27, 1971; Wynkoop, "Panel Asks Removal of Junction Tailings"; Rees, "Committee Recommends Uranium Tailings Action"; Tom Rees, "EPA Denies Two Officials' Views on Tailings Is 'Policy'"; *Rocky Mountain News*, September 29, 1971.

44. *Tailings for Construction Hearing*, 7, 9, 56, 78–79, 101–2.

45. Ibid., 10, 78–79; Joseph A. Lieberman to Wayne Aspinall, October 9, 1971, Box 7822, AEC Secretariat Papers.

46. *Tailings for Construction Hearing*, 101–2; Wynkoop, "Panel Asks Removal of Junction Tailings"; Erlewine to Schlesinger, September 29, 1971.

47. *Tailings for Construction Hearing*, 2, 121; Leonard Larsen, "Dr. Cleere Target of Hostile Probing," *Denver Post*, October 31, 1971, 14.

48. Ripley, "City in Colorado Awakens to Scope of Radioactive Waste Problem"; Richard D. Lyons, "Radiation Danger in Debris Discounted," *New York Times*, October 29, 1971; *Tailings for Construction Hearing*, 14–15, 235–70; Robert D. O'Neill to Edward J. Bauser, October 21, 1971, Box 662, JCAE Papers.

49. *Tailings for Construction Hearing*, 16, 24; Greenleigh to Legal Files, December 18, 1970.

50. Director, AEC Division of Licensing and Regulation, to Edward G. Littel, September 1, 1959; Jay B. Bell and Richard E. Turley, *The Need for Remedial Action and Federal Participation in the Case of the Abandoned Vitro Uranium-Mill Tailings Located in Salt Lake County, Utah* (Salt Lake City: Office of the State Science Advisor, March 7, 1974), 5–6; George W. Soffe to M. J. Pescor, April 18, 1960; James D. Wharton to M. J. Pescor, July 14, 1961, and James D. Wharton to Luther L. Terry, July 26, 1961; all above are located in Utah Department of Environmental Quality, Vitro Case Records, Salt Lake City, Utah. *Uranium Mill Tailings Control Hearing*, 57.
51. *Tailings for Construction Hearing*, 5, 15–16; L. K. Olson, memorandum to Harold L. Price, April 15, 1960, Accession No. 9210120199, NRC Papers; Greenleigh to Legal Files, December 18, 1970; Neil D. Naiden, memorandum to Harold L. Price, December 7, 1960, Accession No. 9210130222, NRC Papers; Joint Committee on Atomic Energy, *Hearings Before the Joint Committee on Atomic Energy on Federal-State Relationships in the Atomic Energy Field*, 86th Cong., 1st. sess., May 19–22, August 26, 1959, 322 (hereafter *Federal-State Relationships Hearing*).
52. U.S.C. sec. 2021 (a)(3) (1970).
53. The Atomic Energy Act was amended in 1959 to add section 274, titled "Cooperation with States," which empowered the AEC to cede many of its regulatory powers to the states. In order for a state to assume the AEC's functions, the governor of the state had to enter into an agreement with the AEC detailing the extent of the transfer of authority. 42 U.S.C. sec. 2021, et sec. Joint Committee on Atomic Energy, *Selected Materials on Federal-State Cooperation in the Atomic Energy Field* (Washington, D.C.: Government Printing Office, 1959), 32; Renee Baruch and Madonna Ghandi, "Radioactive Waste: A Failure in Governmental Regulation," *Albany Law Review* 37 (1972): 127. "Unimportant" was defined by AEC regulation as materials containing less than 0.05 percent by weight uranium or thorium. *Tailings for Construction Hearing*, 19–20; 10 CFR 40.139(a)(1972).
54. Agreement, Article 4.
55. *U.S. Code Congressional and Administrative News* 2 (1959): 2872; 42 U.S.C. sec 2021; Baruch and Ghandi, "Radioactive Waste," 125–34; *Tailings for Construction Hearing*, 107–8, 113–17, 121; *Federal-State Relationships Hearing*, 307–8.
56. Rapoport, "The Trouble with 90.5 Million Tons of Radioactive Tailings"; Baruch and Ghandi, "Radioactive Waste," 100; *Tailings for Construction Hearing*, 102–5, 139–42, 148, 282.
57. Ripley, "Radioactive Building Sand Stirs Dispute"; *Tailings for Construction Hearing*, 103–7; 42 U.S.C. sec. 2014(e); *Federal-State Relationships Hearing*, 322.
58. *Tailings for Construction Hearing*, 24–45, 108, 400–403; Donald A. Nussbaumer, memorandum to Stanley T. Robinson, March 6, 1970, Accession No. 9210120430, NRC Papers; Lyall Johnson, memorandum to Harold L. Price, August 24, 1970, Accession No. 9210120426, NRC Papers; Edward J. Bauser to Martin B. Biles, July 22, 1971, Accession No. 9210150093, NRC Papers; Biles to Erlewine, July 23, 1971, Accession No. 9210120135, NRC Papers; C. L. Henderson, memorandum to Harold L. Price, August 5, 1971, Accession No. 9210120141, NRC Papers; "Stella A.", memorandum to Price, August 17, 1971, Accession No. 9210120154, NRC Papers; McGinley to Youngberg, August 31, 1971, Accession No. 9210120222, NRC Papers; Biles to Bauser, September 28, 1971, Accession No. 9210120230, NRC Papers; Edward J. Bauser, memorandum to Wayne Aspinall, September 28, 1971, Box 426, Aspinall Papers; Grand Junction *Daily Sentinel*, September 15, 1970. *Tailings for Construction Hearing*, 24–44, 108, 395, 400–403; Metzger, *Atomic Establishment*, 181–83.

59. McBride to Price, March 27, 1969; McBride to Price, March 11, 1969; Martin B. Biles, memorandum to Rafford L. Faulkner, March 10, 1969, Accession No. 9210120386, NRC Papers *Tailings for Construction Hearing*, 6–7, 42–43; Ernest C. Tsivoglou and R. L. O'Connell, *Waste Guide for the Uranium Milling Industry* (Cincinnati: Robert A. Taft Engineering Center, 1962), 69; "U.S. Atomic Energy Commission Statement," October 28–29, 1971, Box 426, Aspinall Papers, 3–4; "AEC Participation in Uranium Tailings Remedial Action" (undated) (SECY-1806, June 29, 1971), Box 7822, AEC Secretariat Papers; Harold D. Anamosa, "Proposed Revision of 10 CFR Part 40, 'Licensing of Source Material,'" July 22, 1960 (AEC-R 18/2, 22, 1960), Accession No. 9210120308, NRC Papers; Johnson to Price, August 24, 1970.
60. Anthony Ripley, "Did Someone Make a Deadly Blunder?" *New York Times* (Week in Review), October 3, 1971; *Rocky Mountain News*, September 29, 1971.
61. *Tailings for Construction Hearing*, 101, 117.
62. Ibid., 105.
63. Ibid., 121–22; Anthony Ripley, "Radioactive Sands Linked to Higher Death Rates," *New York Times*, October 28, 1971; Lyons, "Radiation Danger in Debris Discounted"; Metzger, "'Dear Sir'"; "JCAE Inquiry into Mill Tailings Controversy Is Inconclusive," *Nuclear Industry* 18 (October–November, 1971): 31–39; Peter Dominick to James Schlesinger, September 28, 1971, Box 7822, AEC Secretariat Papers.
64. *Science News* 100 (December 11, 1971): 390; *Time* 98 (December 20, 1971): 56; Wayne Aspinall to James Schlesinger, December 2, 1971, Box 7822, AEC Secretariat Papers; James Schlesinger to John O. Pastore, May 16, 1972, Box 7822, AEC Secretariat File; James Schlesinger, draft letter to George P. Shultz, and draft letter to President of the Senate and Speaker of the House (both undated) (SECY-2359, March 13, 1972), Box 7822, AEC Secretariat Papers.
65. *Denver Post*, September 16, 1971; John O. Pastore and Melvin Price to James Schlesinger, February 1, 1972 (SECY-2359, March 13, 1972), Box 7822, AEC Secretariat Papers; Congress, House, *A Bill to Provide Federal Financial Assistance to Limit Radiation Exposure Resulting from the Use of Uranium Mill Tailings in the Area of Grand Junction, Colorado*, 92d Cong., 2d sess. (February 9, 1972), H.R. 13068; Congress, Senate, *A Bill to Provide Federal Financial Assistance to Limit Radiation Exposure Resulting from the Use of Uranium Mill Tailings in the Area of Grand Junction, Colorado*, 92d Cong., 2d sess. (February 9, 1972), S. 3150; Martin R. Hoffman, memorandum to James Schlesinger et al., March 10, 1972 (SECY-2359, March 13, 1972), Box 7822, AEC Secretariat Papers; "Comparative Analysis of AEC Proposed Legislation and Dominick-Aspinall Bills" (undated), (SECY-2359, March 13, 1972), Box 7822, AEC Secretariat Papers; Schlesinger to Pastore, May 16, 1972.
66. P.L. No. 92–314, Title II, 86 Stat. 222 (June 16, 1972), and 10 CFR 12, "Grand Junction Remedial Action Criteria"; Hoffman to Schlesinger et al., March 10, 1972; AEC Press Release No. P-339, October 17, 1972, Box 7930, AEC Secretariat Papers. For a detailed review of the Grand Junction remedial program, see U.S. Comptroller General, *Report to the Congress: Controlling the Radiation Hazard from Uranium Mill Tailings* (Washington, D.C.: General Accounting Office, May 1975).
67. Baruch and Ghandi, "Radioactive Waste," 132.

Chapter 7

1. James B. Graham, memorandum to file, August 15, 1972, Box 662, Joint Committee on Atomic Energy General Correspondence File, Record Group 128, National Archives and Records Administration, Washington, D.C. (hereafter JCAE Papers);

U.S. Congress, Joint Committee on Atomic Energy, Subcommittee on Legislation, ERDA *Authorizing Legislation Fiscal Year 1976*, 94th Cong., 1st sess., February 18 and 27, 1975, 1376 (Note: Appendix 3 of this hearing report is U.S. Congress, Joint Committee on Atomic Energy, Subcommittee on Raw Materials, S. 2566 and H.R. 11378: *Uranium Mill Tailings in the State of Utah*, 93d Cong., 2d sess., March 12, 1974) (hereafter *Vitro Hearing*); U.S. Congress, House, Committee on Interior and Insular Affairs, Subcommittee on Energy and the Environment, *Uranium Mill Tailings Control: H.R. 13382, H.R. 12938, H.R. 12535 and H.R. 13049*, 95th Cong., 2d sess., June 26–27, July 10 and 17, 1978, 38 (hereafter *Uranium Mill Tailings Control Hearing*).
2. Martin R. Hoffman to Calvin L. Rampton, February 13, 1973, Box 642, Frank E. Moss Papers, Marriott Library Special Collections, University of Utah, Salt Lake City, Utah (hereafter Moss Papers); Utah Division of Health, Environmental Health Services Branch, "Background Information on Vitro Tailings and Related Problems," June 7, 1978, Utah Department of Environmental Quality, Vitro Case Records, Salt Lake City, Utah (hereafter Utah State Records); Acumenics Research and Technology, Inc., *Document Summary, Previous Ownership and Operational History of the Uranium Processing Site at Salt Lake City, Utah* (prepared for the United States Department of Justice, Land and Natural Resources Division, September 1989), UMTRA Project Office, Albuquerque, N.Mex. The AEC general counsel places the date Vitro first began selling uranium oxide to the AEC at 1949, as did the director of the Utah Division of Health. Ibid; *Vitro Hearing*, 1376; "Statement Prepared by Lyman J. Olsen," March 12, 1974, Box 451, Moss Papers. Other sources confirm that Vitro began selling to the AEC in 1951. "Vitro Mill Tailings: Chronology of Events" (n.d.), Box 58, MX Information Center Papers, Marriott Library Special Collections, University of Utah, Salt Lake City, Utah; *Vitro Hearing*, 1358; "Statement of Dr. James L. Liverman," March 12, 1974, Box 451, Moss Papers. See generally Jay B. Bell and Richard E. Turley, *The Need for Remedial Action and Federal Participation in the Case of the Abandoned Vitro Uranium-Mill Tailings Located in Salt Lake County, Utah* (Salt Lake City: Office of the State Science Advisor, March 7, 1974), Utah State Records.
3. Director, AEC Division of Licensing and Regulation, to Edward G. Littel, September 1, 1959, Utah State Records; *Uranium Mill Tailings Control Hearing*, 57. The AEC approved the transfer of tailings based on 10 CFR, Part 20, sec. 20.302. Bell and Turley, *Need for Remedial Action*, 5–6; George W. Soffe to M. J. Pescor, April 18, 1960, Utah State Records; James D. Wharton to M. J. Pescor, July 14, 1961, Utah State Records; James D. Wharton to Luther L. Terry, July 26, 1961, Utah State Records.
4. *Uranium Mill Tailings Control Hearing*, 60; James Moore to G. D. Carlyle Thompson, September 5, 1967, Utah State Records; *Uranium Mill Tailings Control Hearing*, 57–58; Moore to Thompson, September 5, 1967; Bell and Turley, *Need for Remedial Action*, 1.
5. Calvin L. Rampton to James Schlesinger, November 29, 1972, Utah State Records; Martin R. Hoffman to Calvin L. Rampton, February 13, 1973, Box 642, Moss Papers; and Box 7930, Atomic Energy Commission Secretariat Collection, U.S. Department of Energy Archives, Germantown, Md. (hereafter AEC Secretariat Papers); Calvin L. Rampton to Dixie Lee Ray, July 19, 1973, Box 7930, AEC Secretariat Papers; Wayne Owens to Dixie Lee Ray, August 28, 1973, Box 7930, AEC Secretariat Papers; Dixie Lee Ray to Wayne Owens, September 24, 1973, Box 7930, AEC Secretariat Papers; David Dominick to William Doub, October 1, 1973, Box 7930, AEC Secretariat

Papers; "Extract of SECY-3052, GM Info Report #204 Filed: O&M 6, GM Information Reports," April 16, 1973, Box 7930, AEC Secretariat Papers; John A. Green to Lyman Olsen, March 21, 1973, Utah State Records; Richard F. Turley, memorandum to Utah Nuclear Energy Commission, April 18, 1973, Utah State Records; L. Manning Muntzing to Charles L. Elkins, November 14, 1973, Box 7930, AEC Secretariat Papers.

6. Robert C. Pendleton to Calvin L. Rampton, March 7, 1973, Utah State Records; Lyman J. Olsen to Mike Miller, August 8, 1973, Utah State Records; "Extract of SECY-3052, etc.," April 16, 1973, Box 7930, AEC Secretariat Papers.
7. Bell and Turley, *Need for Remedial Action*; Moore to Thompson, September 5, 1967, Utah State Records. *Vitro Hearing*, 1346, 1361–62, 1371; "Statement of James L. Liverman," March 12, 1974, Box 451, Moss Papers; *Uranium Mill Tailings Control Hearing*, 58; *Vitro Hearing*, 1342–44, 1350–57, 1388–90; Frank E. Moss to Melvin Price, November 1, 1973, Box 642, Moss Papers; Martin B. Biles to Richard Turley, August 30, 1974, Box 662, JCAE Papers; U. S. Department of Energy, *Inactive Uranium Mill Tailing Remedial Action Program: Salt Lake City (Vitro) Site Offsite Decontamination Program Survey, Compilation of Candidate Properties for Remedial Action* (November 18, 1983), UMTRA Project Office, Albuquerque, N.Mex.

In 1978 the Utah State Bureau of Radiation and Occupational Health and the EPA discovered that Firehouse No. 1 had been built on Vitro tailings. Radon measurements showed up to 1.5 WL of radiation. The limit for uranium miners was 0.3 WL, based on a forty-hour workweek. For the general public, exposure was limited to about 10 percent of that radiation exposure, or 0.03 WL. In cases in which exposure was near 24 hours per day, the recommended limit was 0.01 WL. The firemen threatened to strike because of the delays in remedying the problem at the facility. *Uranium Mill Tailings Control Hearing*, 37, 67; Utah Department of Social Services, *News* (press release), June 8, 1978, Utah State Records.
8. Lyman J. Olsen to Calvin L. Rampton, July 16, 1973, Utah State Records; *Uranium Mill Tailings Control Hearing*, 58; John F. O'Leary to Wayne Owens, September 5, 1973, Utah State Records; Lyman J. Olsen, memorandum to "Residents, Employers and Employees in the Vicinity of the Uranium Tailings Pile," July 1, 1975, Utah State Records; Lyman J. Olsen to Harry L. Gibbons, September 22, 1975, Utah State Records; Robert M. Yeates to Don Mackey, November 16, 1973, Utah State Records.
9. S. 2566 and H.R. 11378, 93d Cong., 2d sess., 1974.
10. "JCAE Subcommittee on Raw Materials Announces Hearings on Uranium Mill Tailings Located in the State of Utah" (JCAE Press Release No. 746), February 27, 1974, Box 251, JCAE Papers; *Vitro Hearing*, 1344–50, 1374–79, 1382–86.
11. *Vitro Hearing*, 1375; Richard Turley, memorandum to Mike Miller, March 8, 1974, Utah State Records.
12. *Vitro Hearing*, 1358–59, 1364.
13. Ibid., 1359; H. Peter Metzger, "AEC vs. The Public: The Case of the Uranium Tailings," *Science News* 106 (July 13, 1974): 31.
14. National Environmental Policy Act of 1969, P.L. No. 91–190, 83 Stat. 852.
15. *Vitro Hearing*, 1351–54, 1369–72; R. Tenney Johnson to John O. Pastore, November 7, 1975, Box 530, Moss Papers.
16. *Uranium Mill Tailings Control Hearing*, 1395–97; *Summary Report: Phase I Study of Inactive Uranium Mill Sites and Tailings Piles*, October, 1974, Utah State Records; William Doub to Cecil Andrus, May 3, 1974, Box 7930, AEC Secretariat Papers; William Doub to John Vanderhoof, May 3, 1974, Box 7930, AEC Secretariat Papers;

Russell Train to Alan Bible, April 30, 1974, Box 662, JCAE Papers; Russell Train to Peter Domenici, April 19, 1974, Utah State Records.
17. K. L. DeVries and Jay B. Bell, memorandum to Utah State Advisory Council on Science and Technology, January 2, 1974, Box 642, Moss Papers. The comments made by Smith and McGinley, as well as the data collected by federal and state agencies and individual researchers, about the Vitro tailings led the State Advisory Council to formally throw its support behind Senate Bill 2566.
18. The long-standing controversy about the dual mandates of the AEC, to both promote and regulate atomic energy matters, convinced Congress to restructure the nation's atomic energy bureaucracy. The Energy Reorganization Act of 1974 created two separate organizations, the Energy Research and Development Administration, responsible for promotion and development of atomic energy, and the Nuclear Regulatory Commission, responsible for licensing and safety. The Department of Energy replaced ERDA on October 1, 1977, when DOE became the twelfth cabinet-level department within the federal government.
19. Frank E. Moss to R. Tenney Johnson, November 24, 1975, Box 530, Moss Papers; DeVries and Bell to [Utah] State Advisory Council on Science and Technology, January 2, 1974; *Uranium Mill Tailings Control Hearing*, 70.
20. Dixie Lee Ray to Alan Bible, April 24, 1974, Box 662, JCAE Papers; James L. Liverman to John A. Erlewine, April 11, 1974, Box 7930, AEC Secretariat Papers; "Eight Western Governors Asked to Assist Federal Survey of Uranium Tailings Pile" (AEC Press Release No. 74-14(G)), May 10, 1974, Box 662, JCAE Papers; *Summary Report: Phase I Study of Inactive Uranium Mill Sites and Tailings Piles*, October 1974, Utah State Records.
21. William A. Anders to Frank E. Moss, February 18, 1975, Box 530, Moss Papers; "Summary of AEC Report on Status of Uranium Tailings Piles" (n.d.), *Summary Report: Phase I Study of Inactive Uranium Mill Sites and Tailings Piles*, October 1974, Utah State Records; Russell Train to Frank E. Moss, February 18, 1975, Box 530, Moss Papers; Graham to the files, October 1, 1975; Hal Hollister to Tom Clark, May 4, 1977, Utah State Records.
22. *Uranium Mill Tailings Control Hearing*, 6, 77, 102–4, 111–16, 437–38; Floyd L. Galpin to Richard Kennedy, February 5, 1976, Utah State Records. Comptroller General, U.S. General Accounting Office, *The Uranium Mill Tailings Cleanup: Federal Leadership at Last?* June 20, 1978, 4, reproduced in *Uranium Mill Tailings Control Hearing*, 275–303.
23. "EPA Finds 'Intolerable' Radioactivity in Drinking Water Near Uranium Mines," *Environmental Reporter* 6 (August 22, 1975): 651–52; "Report Says Radon Exposure Major Hazard in Living Near Uranium Tailings Deposits," *Environmental Reporter* 7 (May 7, 1976): 49–50.
24. "NRC Radon Impact Estimates 'Grossly' in Error, Says Ex-Oak Ridge Official," *Environmental Reporter* 8 (November 25, 1977): 1139–40.
25. *Uranium Mill Tailings Control Hearing*, 444–45.
26. Ibid., 76.
27. Ibid., 67, 89.
28. Ibid., 3–35, 46–47, 69–72.
29. Ibid., 111–22.
30. U.S.C. 2021, sec. 274(d) and (g).
31. *Uranium Mill Tailings Control Hearing*, 130, 139–52.
32. Congressional Quarterly, *Weekly Report* 36 (August 19, 1978): 4; P.L. No. 95–604 (November 8, 1978), 92 Stat. 3021 et seq., 42 U.S.C.A. sec. 7901, et seq.

33. U.S.C.A. sec. 7901(a) (1979).
34. U.S. Congress, House Committee on Interior and Insular Affairs, *H.R. Report No. 95–1480*, pt. 2, 95th Cong., 2d sess. (August 11, 1978), 35. There is little definitive legislative history on UMTRCA. Two House reports were submitted, neither of which accurately reflects the language of the final version of the act. See U.S. Congress, House Committee on Interior and Insular Affairs, *H.R. Report No. 95–1480*, pts. 1 and 2, 95th Cong., 2d sess. (August 11, 1978); and U.S. Congress, House Committee on Interstate and Foreign Commerce, *H.R. Report No. 95–1480 (ii)*, 95th Cong., 2d sess. (September 30, 1978).
35. Two sites, Belfield and Bowman, North Dakota, were later added to the UMTRCA list, for a total of twenty-four sites requiring remedial action.
36. U.S.C. sec. 7901–25.
37. National Environmental Policy Act of 1969, P.L. No. 91–190, 83 Stat. 852 (1970).
38. CFR Part 40, Appendix A, criterion 1 (1989).
39. House Committee on Interior and Insular Affairs, *H.R. Report No. 95–1480*, 26. Title III of the act was much more site-specific, directing the NRC to investigate two active mill sites in New Mexico, the former Homestake–New Mexico Partners site near Milan and the Anaconda site near Bluewater, to determine whether either qualified for remedial action under Title I or the regulatory provisions of Title II.
40. Tailings sites at Tuba City and Monument Valley, Arizona; Shiprock, New Mexico; and Mexican Hat, Utah, are all on tribal lands. The Riverton, Wyoming, site is located on private land within the boundaries of the Wind River Indian Reservation (Northern Arapaho and Shoshone Tribes).
41. U.S.C. sec. 7922(a).
42. CFR Part 40.
43. CFR Part 192.
44. National Environmental Policy Act of 1969 (42 U.S.C. sec. 4321, et seq.).
45. Under UMTRCA, "tailings" included the "tailings or wastes from any ore processed primarily for its source material content." 42 U.S.C.A. sec. 2014(e)(2) (1979).
46. The House Committee on Interior and Insular Affairs recognized that the provisions of Title II would put significant financial burdens on the industry. The committee nevertheless "[believed] and [expected] that these purposes should be met without causing mill closings putting people out of work [sic]." It cautioned the NRC to take into account the problem of balancing regulatory compliance with economic hardship and directed it to provide "a means to alleviate or mitigate those problems where appropriate while assuring that the purposes of the act are fully met." The act provided no guidance to the NRC for carrying out this ambivalent mandate. House Committee on Interior and Insular Affairs, *H.R. Report No. 95–1480*, pt. 2 (August 11, 1978), 44.

    One contemporary example of this problem is the 11-million-ton Atlas Minerals Co. tailings pile near Moab, Utah. For nearly forty years, the pile has been leaching chemical and radioactive waste into the Colorado River aquifer, threatening the health of 20 million people who live downstream. Relocating the pile away from the banks of the Colorado River would cost an estimated $155 million, while capping the pile in place would cost $19 million. Atlas is bonded for just $6 million. In fall 1998 Atlas filed for Chapter 11 bankruptcy protection and faced the prospect of complete liquidation if forced to bear the full cost of remedial action. "Parental Care For Uranium Tailings Only Goes So Far," *High Country News*, October 3, 1994; "Groups Launch 2nd Legal Action Against Leaving Tailings by River," *Salt Lake Tribune*, October 14, 1998; "Suits Filed Against Radioactive Flow into Colorado River," *Las Vegas Sun*, October 23, 1998;

"County, Environmentalists Sue Agency Over Tailings," *Salt Lake Tribune*, November 12, 1998; "Babbitt OK with Moving Uranium Mill's Wastes," *Deseret News*, November 24, 1998.

47. It should be noted that states may lack objectivity in matters of health and environmental protection, and federal regulators are less subject to local political pressure than are their state counterparts. Most important, most states simply did not have the funds or technical expertise to undertake remedial action programs or adequately manage the tailings for the long-term future. See John Magee, "The Uranium Mill Tailings Radiation Control Act of 1978," *Ecology Law Quarterly* 8 (1980): 801–9; Elisa J. Grammer, "The Uranium Mill Tailings Radiation Control Act of 1978 and NRC's Agreement State Program," *Natural Resources Lawyer* 13 (1981): 469–522; and "Regulation of Uranium Mill Tailings; Uranium Mill Tailings Radiation Control Act of 1978," in Frank P. Grad, ed., *Treatise on Environmental Law* (New York: Matthew Bender & Co., 1992), 4:6–124–6 to 134.

Chapter 8

1. Holger Albrethsen, Jr., and Frank E. McGinley, *Summary History of Domestic Uranium Procurement under U.S. Atomic Energy Commission Contracts* (Grand Junction, Colo.: USDOE Area Field Office, September 1982), 4–5. Under the terms of the last stretch-out program (1966–70), the AEC allowed uranium mills to delay delivery of portions of their yellowcake beyond the 1966 deadline. The AEC agreed to purchase quantities of yellowcake in 1969 and 1970 equal to the amounts already deferred.
2. See generally George Dannenbaum, *Boom to Bust: Remembrances of the Grants, New Mexico Uranium Boom* (Albuquerque: Creative Designs, 1994).
3. Comptroller General, U.S. General Accounting Office, *The Uranium Mill Tailings Cleanup: Federal Leadership at Last?* June 20, 1978, 21–24; reproduced in U.S. Congress, House, Committee on Interior and Insular Affairs, Subcommittee on Energy and the Environment, *Uranium Mill Tailings Control: H.R. 13382, H.R. 12938, H.R. 12535 and H.R. 13049*, 95th Cong., 2d sess., June 26–27, July 10, 17, 1978, 214–25 (hereafter *Uranium Mill Tailings Control Hearing*).
4. Ibid.
5. U.S. Department of Energy, *Integrated Data Base for 1988: Spent Fuel and Radioactive Waste Inventories, Projections, and Characteristics*, DOE/RW-0006 (Washington, D.C.: Oak Ridge National Laboratory, September 1988), 156; Luther J. Carter, *Nuclear Imperatives and Public Trust* (Washington, D.C.: Resources for the Future, 1987), 13; Comptroller General, *Uranium Mill Tailings Cleanup*, 24.
6. Chris Shuey, "Bringing Tailings under Control," *Workbook* 10 (1985): 110–11; Uranium Licensing Requirements, 45 Fed. Reg. 65,521–65,528 (1980). See generally 10 CFR 40, Appendix A (1995). In fact, 200 or 1,000 years was hardly material because EPA figures demonstrated that the tailings needed to be stored 265,000 years to reduce radon 222 emissions to 10 percent of their initial levels. Office of Radiation Programs, U.S. Environmental Protection Agency, *Final Rule for Radon-222 Emissions from Licensed Uranium Mill Tailings* (Background Information Paper, August 1986).
7. Mark Crawford, "Mill Tailings: A $4 Billion Problem," *Science* 229 (August 9, 1985): 537–38.
8. Marian Radetzki, *Uranium: A Strategic Source of Energy* (New York: St. Martin's Press, 1981), 97; "U$_3$O$_8$: Energy from Wyoming's Powerful Sand," *In Wyoming* 13 (February–March 1980): 49.
9. See generally Norman Moss, *The Politics of Uranium* (New York: Universe Books, 1982);

June H. Taylor and Michael D. Archil, *Yellowcake* (New York: Pergamon Press, 1979); Earle Gray, *The Great Uranium Cartel* (Toronto: McClelland and Stewart, 1982).

10. Radetzki, *Uranium*, 87–94; Michael A. Amundson, "Home on the Range No More: The Boom and Bust of a Wyoming Uranium Mining Town," *Western Historical Quarterly* 26 (1995): 497–98.

11. "New Mexico's Uranium Mine Tailings Problem and the Integrated Costs Equation," June 1978, Southwest Research Information Center Archives, Albuquerque, N.Mex.; Federal Energy Administration, *New Mexico Uranium, 1950–2000* (Dallas, Tex.: FEA Region VI Office, December 1976), 46; Anthony D. Owen, *Economics of Uranium* (New York: Praeger, 1985), 39–42; Amundson, "Home on the Range No More," 491–92; John E. Chubb, *Interest Groups and the Bureaucracy: The Politics of Energy* (Stanford: Stanford University Press, 1983), 96; Office of Nuclear Material Safety and Safeguards, U.S. Nuclear Regulatory Commission, *Draft Generic Environmental Impact Statement on Uranium Milling* (April 1979), 3-1–3-14.

12. Crawford, "Mill Tailings," 537–38; Anthony D. Owen, "Short-Term Price Formation in the U.S. Uranium Market," *Energy Journal* 6 (1985): 37. See generally U.S. Congress, Senate Committee on Energy and Natural Resources, *Status of the Domestic Uranium Mining and Milling Industry: Hearing on the Effects of Imports Before the Subcommittee on Energy Research and Development of the Senate Committee on Energy and Natural Resources*, 97th Cong., 1st sess., 1981 (testimony of George B. Rice, American Mining Congress); U.S. Congress, House Committee on Armed Services, *Management of Commingled Uranium Mill Tailings: Hearings Before the Procurement and Military Nuclear Systems Subcommittee of the House Committee on Armed Services*, 97th Cong., 2d sess., 1982 (testimony of George B. Rice and Jack Vogt, American Mining Congress); U.S. Department of Energy, Energy Information Administration, *Domestic Uranium Mining Industry 1991 Viability Assessment*, DOE/EIA-0477(91) (Washington, D.C.: USDOE, 1992).

13. Energy and Water Development Appropriation Act of 1982, P.L. No. 97–88, Title IV, 95 Stat. 1135, 1147–48; Nuclear Regulatory Commission Appropriations Authorization, P.L. No. 97–415, 96 Stat. 2067, 2077–78.

14. Shuey, "Bringing Tailings under Control," 110–13; Crawford, "Mill Tailings," 537; U.S. Congress, Joint Committee on Atomic Energy, *Hearings Before the Special Subcommittee on Radiation of the Joint Committee on Atomic Energy*, 86th Cong., 1st sess., 1959, 684–90; Ken Silver, "The Yellowed Archives of Yellowcake," *Public Health Reports* 111 (March–April 1996), 124; Albrethsen and McGinley, *Summary History*, B-6.

15. Kerr-McGee Nuclear Corp. v. N.R.C., 17 Envt'l. Rep. Cas. (BNA) 1537 (1982); Elizabeth V. Scott, "Unfinished Business: The Regulation of Uranium Mining and Milling," 18 *University of Richmond Law Review* 615 (1984).

16. John D. Collins, "Uranium Mine and Mill Tailings Reclamation in Wyoming: Ten Years after the Industry Collapsed," *Land and Water Law Review* 26 (1991): 505–6; American Mining Congress v. Thomas, 772 F.2d 640 (10th Cir., 1985), *cert. denied* 479 U.S. 814 (1986).

17. Atlas v. United States, 15 Cl. Ct. 681 (1988), aff'd 895 F.2d 745 (Fed. Cir. 1990); "Appellate Court Upholds Uranium Mill Tailings Act," *Albuquerque Journal*, January 31, 1989; Mary Boaz, "Retroactive Liability for Clean-up of Hazardous Waste in *Atlas v. United States*: The Nuclear Industry's Failed Attempt to Make the Government Pay," 6 *Journal of Mineral and Law Policy* 275 (1990–91). In addition to Atlas, the plaintiffs were Kerr-McGee Chemical Corp., Quivira Mining Co., Western Nuclear,

Inc., Atlantic Richfield Co., Umetco Minerals Corp., Union Carbide Corp., Homestake Mining Co. of California, Inc., and Pathfinder Mines Corp.

18. UMTRA: *Fiscal Year 1995 Annual Report to Stakeholders*; Crawford, "Mill Tailings," 537.
19. Crawford, "Mill Tailings," 537–38. For an example of bankruptcy associated with tailings management, see the problem of the Atlas mill, located near Moab. Atlas filed Chapter 11 bankruptcy protection in 1998 to shield it from the ruinous cost of stabilizing its pile, estimated to cost over $6 million if capped in place, or as much as $120 million if the tailings are removed. "Babbitt OK with Moving Uranium Mill's Wastes," *Deseret News*, November 24, 1998; "Groups Threaten to Sue Feds Over Mill-Tailings Pile," Grand Junction *Daily Sentinel*, November 11, 1998; "Some Fear That Colorado Is Getting Nuked," *High Country News*, May 26, 1997; "Parental Care for Uranium Tailings Only Goes So Far," *High Country News*, October 3, 1994.
20. CFR Part 192, "Standards for Remedial Actions at Inactive Uranium Processing Sites" (effective March 7, 1983). The agency was sued by environmentalists, including the Environmental Defense Fund, the Natural Resources Defense Council, and the Sierra Club; and by the American Mining Congress, representing industry. The environmental litigants alleged that EPA standards for radon were too permissive and that they failed to adequately protect groundwater supplies. Shuey, "Bringing Tailings under Control," 110–13.
21. Collins, "Uranium Mine and Mill Tailings Reclamation in Wyoming," 23–98.
22. "Memorandum of Understanding, EPA and NRC," 56 Fed. Reg. 55,434 (1991).
23. The Maybell and Naturita sites were licensed by the NRC in 1999.
24. Energy Policy Act of 1992, P.L. No. 102–486, 106 Stat 2776 (1992).
25. W. Kip Viscusi, *Fatal Tradeoffs: Public and Private Responsibilities for Risk* (New York: Oxford University Press, 1992), 52. In comparison, Viscusi estimates that regulating grain dust saves four lives per year at $5.3 million per life; regulating uranium mines saves 1.1 lives per year at $6.9 million per life; benzene restrictions save 3.8 lives per year at $17.1 million per life; regulating glass plants saves .11 lives per year at $19.2 million; regulating copper smelters saves .06 lives per year at $26.5 million per life; regulating ethylene oxide saves 2.8 lives annually at $25.6 million per life; regulating low-arsenic copper saves .09 lives per year at $764 million per life; regulating land disposal facilities saves 2.52 lives per year at $3.5 billion per life; and formaldehyde controls save .01 lives per year at $72,000 million per life. All figures are expressed in 1984 dollars.

Bibliography

Books

Adato, Michelle, James MacKenzie, Robert Pollard, and Ellyn Weiss. *Safety Second: The NRC and America's Nuclear Power Plants*. Bloomington: Indiana University Press, 1987.
Allardice, Corbin, and Edward R. Trapnell. *The Atomic Energy Commission*. New York: Praeger, 1974.
Arrington, Leonard J., and Anthony T. Cluff. *Federally Financed Industrial Plants Constructed in Utah during World War II*. Logan: Utah State University Press, 1969.
Ball, Howard. *Cancer Factories: America's Tragic Quest for Uranium Self-Sufficiency*. Westport, Conn.: Greenwood Press, 1993.
———. *Justice Downwind: America's Atomic Testing Program in the 1950s*. Oxford: Oxford University Press, 1986.
Balogh, Brian. *Chain Reaction: Expert Debate and Public Participation in American Commercial Nuclear Power, 1945–1975*. Cambridge: Cambridge University Press, 1991.
Bartlett, Donald, and James Steele. *Forevermore: Nuclear Waste in America*. New York: Norton, 1985.
Bertell, Rosalie. *No Immediate Danger: Prognosis for a Radioactive Earth*. London: Women's Press Limited, 1985.
Boyer, Paul. *By the Bomb's Early Light: American Thought and Culture at the Dawn of the Atomic Age*. New York: Pantheon Books, 1985.
Brown, Corinne, and Robert Monroe. *Time Bomb: Understanding the Threat of Nuclear Power*. New York: William Morrow, 1981.
Bruyn, Kathleen. *Uranium Country*. Boulder: University of Colorado Press, 1955.
Carter, Luther J. *Nuclear Imperatives and the Public Trust*. Washington, D.C.: Resources for the Future, 1987.
Chubb, John E. *Interest Groups and the Bureaucracy: The Politics of Energy*. Stanford: Stanford University Press, 1983.
Clarfield, Gerald H., and William M. Wiecek. *Nuclear America: Military and Civilian Nuclear Power in the United States 1940–1980*. New York: Harper and Row, 1984.

Cohn, Steven Mark. *Too Cheap to Meter: An Economic and Philosophical Analysis of the Nuclear Dream*. Albany: State University of New York Press, 1997.
Compton, Arthur H. *Atomic Quest: A Personal Narrative*. New York: Oxford University Press, 1956.
Curie, Eve. *Madam Curie*. Garden City, N.Y.: Doubleday, Doran, 1938.
Dannenbaum, George. *Boom to Bust: Remembrances of the Grants, New Mexico Uranium Boom*. Albuquerque, N.Mex.: Creative Designs, 1994.
Dawson, Frank G. *Nuclear Power: Development and Management of a Technology*. Seattle: University of Washington Press, 1976.
Del Sesto, Steven L. *Science, Politics, and Controversy: Civilian Nuclear Power in the United States, 1946–1974*. Boulder, Colo.: Westview Press, 1979.
DeMent, Jack, and H. C. Drake. *Uranium and Atomic Power*. Brooklyn, N.Y.: Chemical Publishing Company, 1941.
Edelstein, Michael R., and William J. Makofske. *Radon's Deadly Daughters: Science, Environmental Policy and the Politics of Risk*. Lanham, Md.: Rowman and Littlefield, 1998.
Ford, Daniel. *The Cult of the Atom: The Secret Papers of the Atomic Energy Commission*. New York: Simon and Schuster, 1982.
Gomez, Arthur R. *Quest for the Golden Circle: The Four Corners and the Metropolitan West, 1945–1970*. Albuquerque: University of New Mexico Press, 1994.
Goodchild, Peter. *J. Robert Oppenheimer: "Shatterer of Worlds."* London: BBC, 1980.
Gray, Earle. *The Great Uranium Cartel*. Toronto: McClelland and Stewart, 1982.
Green, Harold P., and Alan Rosenthal. *Government of the Atom*. New York: Atherton Press, 1963.
Grigg, E. R. N. *The Trail of the Invisible Light*. Springfield, Ill: Charles C. Thomas, 1965.
Groves, Leslie M. *Now It Can Be Told: The Story of the Manhattan Project*. New York: Da Capo Press, 1963.
Gwin, Louis. *Speak No Evil: The Promotional Heritage of Nuclear Risk Communication*. New York: Praeger, 1990.
Hays, Samuel P. *Beauty, Health, and Permanence: Environmental Politics in the United States, 1955–1985*. Cambridge: Cambridge University Press, 1987.
Helmreich, Jonathan E. *Gathering Rare Ores: The Diplomacy of Uranium Acquisition, 1943–1954*. Princeton: Princeton University Press, 1986.
Hersey, John. *Hiroshima*. New York: Alfred A. Knopf, 1946.
Hertsgaard, Mark. *Nuclear, Inc.: The Men and Money Behind the Nuclear Power Industry*. New York: Pantheon Books, 1983.
Hewlett, Richard G., and Oscar E. Anderson, Jr. *The New World, 1939–1946: A History of the United States Atomic Energy Commission*. University Park: Pennsylvania State University Press, 1962.
Hewlett, Richard G., and Francis Duncan. *Atomic Shield, 1947–1952: A History of the United States Atomic Energy Commission*. University Park: Pennsylvania State University Press, 1969.
Hilgartner, Stephen, Richard C. Bell, and Rory O'Connor. *Nukespeak: Nuclear Language, Visions, and Mindset*. San Francisco: Sierra Club Books, 1982.
Hueper, W. C. *Occupational and Environmental Cancer of the Respiratory System*. Springfield, Ill.: Charles C. Thomas, 1942.
Jasper, James M. *Nuclear Politics: Energy and the State in the United States, Sweden, and France*. Princeton: Princeton University Press, 1990.

Katz, Joseph J., and Eugene Rabinowitch. *The Chemistry of Uranium (Part 1): The Element, Its Binary and Related Compounds*. New York: McGraw-Hill, 1951.

Keener, James, and Christine Bebee Keener. *Colorado Highway 141, Unaweep to Uravan: Travel through 1.7 Billion Years into the Atomic Age*. Grand Junction, Colo.: Grand River Publishing, 1988.

Kline, Dan, and Ward Lloyd, eds. *The History of Glass*. London: Orbis, 1984.

Kneese, Allen V., and Charles L. Schultze. *Pollution, Prices, and Public Policy*. Washington, D.C.: Brookings Institution Press, 1975.

Lewis, H. W. *Technological Risk*. New York: Norton, 1990.

Lewis, Richard S. *The Nuclear-Power Rebellion*. New York: Viking Press, 1972.

Lilienthal, David E. *Change, Hope and the Bomb*. Princeton: Princeton University Press, 1963.

———. *The Journals of David Lilienthal*, vol. 2, *Atomic Energy Years, 1945–1950*. New York: Harper and Row, 1964.

Mazuzan, George T., and J. Samuel Walker. *Controlling the Atom: The Beginnings of Nuclear Regulation, 1946–1962*. Berkeley: University of California Press, 1984.

Merritt, Robert C. *The Extractive Metallurgy of Uranium*. Golden: Colorado School of Mines Research Institute, 1971.

Metzger, H. Peter. *The Atomic Establishment*. New York: Simon and Schuster, 1972.

Miller, James. *Democracy in the Streets*. New York: Simon and Schuster, 1987.

Moss, Norman. *The Politics of Uranium*. New York: Universe Books, 1982.

Mullenbach, Philip. *Civilian Nuclear Power*. New York: Twentieth Century Fund, 1963.

Nader, Ralph. *The Menace of Atomic Energy*. New York: Norton, 1979.

Nash, Gerald. *The American West in the Twentieth Century: A Short History of an Urban Oasis*. Englewood Cliffs, N.J.: Prentice-Hall, 1973.

Newell, Maxine. *Charlie Steen's Mi Vida*. Moab, Utah: Moab's Printing Place, 1992.

Newman, James R., and Byron S. Miller. *Control of Atomic Energy: A Study of Its Social, Economic, and Political Implications*. New York: McGraw-Hill, 1948.

Nininger, Robert D. *Minerals for Atomic Energy*. New York: D. Van Nostrand, 1954.

Nordlinger, Eric A. *On the Autonomy of the Democratic State*. Cambridge, Mass.: Harvard University Press, 1981.

Orlans, Harold. *Contracting for Atoms*. Washington, D.C.: Brookings Institution Press, 1967.

Owen, Anthony D. *Economics of Uranium*. New York: Praeger, 1985.

Radetzki, Marian. *Uranium: A Strategic Source of Energy*. New York: St. Martin's Press, 1981.

Ringholz, Raye C. *Uranium Frenzy: Boom and Bust on the Colorado Plateau*. Albuquerque: University of New Mexico Press, 1991.

Roberts, Lewis E. J. *Nuclear Power and Public Responsibility*. Cambridge: Cambridge University Press, 1984.

Rolph, Elizabeth S. *Nuclear Power and Public Safety: A Study in Regulation*. Lexington, Mass.: D. C. Heath, 1979.

Romer, Alfred, ed. *The Discovery of Radioactivity and Transmutation*. New York: Dover Publications, 1964.

Schubert, Jack, and Ralph E. Lapp. *Radiation: What It Is and How It Affects You*. New York: Viking Press, 1957.

Schurr, Sam H., and Jacob Marshank. *Economic Aspects of Atomic Power*. Princeton: Princeton University Press, 1950.

Schwartz, Stephen I. *Atomic Audit: The Costs and Consequences of U.S. Nuclear Weapons since 1940*. Washington, D.C.: Brookings Institution Press, 1998.
Simon, Herbert A. *Reason in Human Affairs*. Stanford: Stanford University Press, 1983.
Smyth, Henry DeWolf. *Atomic Energy for Military Purposes: The Official Report on the Development of the Atomic Bomb under the Auspices of the United States Government, 1940–1945*. Princeton: Princeton University Press, 1945.
Soddy, Frederick. *Interpretation of Radium*. London: John Murray, 1909.
Taylor, June H., and Michael D. Yorkell. *Yellowcake*. New York: Pergamon Press, 1979.
Taylor, Raymond W., and Samuel W. Taylor. *Uranium Fever, or No Talk under $1 Million*. New York: Macmillan, 1970.
Teller, Edward, and Alan Brown. *The Legacy of Hiroshima*. New York: Doubleday, 1962.
Thomas, Morgan. *Atomic Energy and Congress*. Ann Arbor: University of Michigan Press, 1956.
United States Vanadium Company. *Mesa Miracle in Colorado, Utah, New Mexico, Arizona*. New York: United States Vanadium Company, Union Carbide and Carbon Corporation, 1952.
Viscusi, W. Kip. *Fatal Tradeoffs: Public and Private Responsibilities for Risk*. New York: Oxford University Press, 1992.
Walker, J. Samuel. *Containing the Atom: Nuclear Regulation in a Changing Environment, 1963–1971*. Berkeley: University of California Press, 1992.
Wells, H. G. *The World Set Free: A Story of Mankind*. London: Macmillan, 1914.

Government Documents

Acumenics Research and Technology, Inc. *Document Summary, Previous Ownership and Operational History of the Uranium Processing Site at Salt Lake City, Utah*. Prepared for the U.S. Department of Justice, Land and Natural Resources Division, September 1989.
Albrethsen, Holger, Jr., and Frank E. McGinley. *Summary History of Domestic Uranium Procurement under U.S. Atomic Energy Commission Contracts*. Grand Junction, Colo.: USDOE Area Field Office, September 1982.
Bell, Jay B., and Richard E. Turley. *The Need for Remedial Action and Federal Participation in the Case of The Abandoned Vitro Uranium-Mill Tailings Located in Salt Lake County, Utah*. Salt Lake City: Office of the State Science Advisor, March 7, 1974.
Bolton, H. Carrington. "Index to the Literature of Uranium, 1789–1885." *Annual Report of the Board of Regents of the Smithsonian Institution, Part I (to July, 1885)*. Washington, D.C.: Government Printing Office, 1886.
Boutwell, J. M. "Vanadium and Uranium in Southeastern Utah." *United States Geological Survey Bulletin 260*. Washington, D.C.: Government Printing Office, 1905.
Cohen, J. B., et al. *Waste Characteristics for the Acid-Leach Solvent Extraction Uranium Refining Process, II, Climax Uranium Company*. USPHS Technical Report W62–17. Robert A. Taft Sanitary Engineering Center, Cincinnati, Ohio, 1962.
———. *Waste Characteristics for the Carbonate-Leach Uranium Extraction Process, I, Homestake–New Mexico Partners Company*. USPHS Technical Report W62–17. Robert A. Taft Sanitary Engineering Center, Cincinnati, Ohio, 1962.
Colorado Department of Public Health. *Uranium Wastes and Colorado's Environment*, 2d ed. Denver: Colorado Department of Public Health, 1971.
Congressional Quarterly. *Weekly Report* 36 (August 19, 1978): 4.
Congressional Record 92 (1946): 6097.

Congressional Record 116 (pt. 21) (August 12, 1970): 28,579–81.
Environmental Standards for Uranium and Thorium Mill Tailings at Licensed Commercial Processing Sites. 48 Fed. Reg. 45 (1983): 45,926–28.
Federal Energy Administration. *New Mexico Uranium, 1950–2000*. Dallas, Tex.: FEA Region VI Office, 1976.
Federal Radiation Council. *Radiation Protection Guidance for Federal Agencies*. Memorandum for the President, May 13, 1960.
Fleck, Herman, and William G. Haldane, "A Study of the Uranium and Vanadium Belts of Southern Colorado." *Report of the Colorado State Bureau of Mines for the Years 1905–6*. Denver: Colorado State Bureau of Mines, 1907. 47–115.
George, R. D. *Common Minerals and Rocks: Their Occurrence and Uses (Colorado Geological Survey Bulletin 12)*. Denver: Eames Bros., State Printers, 1917.
Gillett, H. W., and E. L. Mack. "Preparation of Ferro-uranium." *United States Bureau of Mines Technical Paper 177*. Washington, D.C.: Government Printing Office, 1917.
Glauberman, H., and A. J. Breslin. *Health and Safety Laboratory Report HASL 64-14*. United States Atomic Energy Commission, New York Operations Office, July 31, 1964.
Harris, W. B., et al. *Environmental Hazards Associated with the Milling of Uranium Ore, HASL-40*. United States Atomic Energy Commission, Health and Safety Laboratory, New York Operations Office, June 1958.
Lush, D., et al. "An Assessment of the Long Term Interaction of Uranium Tailings with the Natural Environment." In *Proceedings of the Seminar on Management, Stabilization and Environmental Impact of Uranium Mill Tailings*. Albuquerque, N.Mex.: OECD Nuclear Energy Agency, July 1978.
Malcolm, Wyatt. *Notes on Radium-Bearing Minerals (Canada Department of Mines, Prospector's Handbook No. 1)*. Ottawa: Government Printing Office, 1914.
Matignon, Camille. "The Manufacture of Radium." *Annual Report of the Board of Regents of the Smithsonian Institution, 1925*. Washington, D.C.: Government Printing Office, 1926.
Miller, Arthur L. "Personal Reminiscences of the Early History of the Radium Extraction Industry in the U.S.A." *Argonne National Laboratory Report ANL-7461* (July 1968): 91–100.
Moore, Richard B., and Karl L. Kithil. "A Preliminary Report on Uranium, Radium, and Vanadium." *United States Bureau of Mines, Bulletin 70*. Washington, D.C.: Government Printing Office, 1913.
National Bureau of Standards. *Maximum Permissible Body Burdens and Maximum Permissible Concentrations of Radionuclides in Air and Water for Occupational Exposure (Handbook 69)*. Washington, D.C.: Government Printing Office, June 1959.
National Lead Company (for USAEC). *Winchester Laboratory Report WIN-112*. February 1, 1960.
———. *Winchester Laboratory Report WIN-125*. September 30, 1961.
National Research Council, Committee on the Biological Effects of Ionizing Radiation. *Health Risks of Radon and Other Internally Deposited Alpha-Emitters, Beir IV*. 1988.
Pahren, J. R., et al. *Waste Characteristics for the Carbonate-Leach Uranium Extraction Process, II, Homestake-Sapin Company*. USPHS Technical Report W62–17. Robert A. Taft Sanitary Engineering Center, Cincinnati, Ohio, 1962.
Perkins, Betty L. *An Overview of the New Mexico Uranium Industry*. Santa Fe: New Mexico Energy and Minerals Department, January 1979.

Pratt, Joseph Hyde. "Mineral Resources." *Twenty-first Annual Report of the United States Geological Survey to the Secretary of the Interior 1899–1900*. Washington, D.C.: Government Printing Office, 1901.

Shearer, S. D., et al. *Waste Characteristics for the Acid-Leach Solvent Extraction Uranium Refining Process, I, Gunnison Mining Company*. USPHS Technical Report W62–17. Robert A. Taft Sanitary Engineering Center, Cincinnati, Ohio, 1962.

Snelling, Robert N., and Robert D. Siek. *Evaluation of Radon Film Badge*. Southwestern Radiological Health Laboratories, USPHS and Colorado Department of Public Health, 1968.

Swift, J. J., J. M. Hardin, and H. W. Calley. *Potential Radiological Impact of Airborne Releases and Direct Gamma Radiation to Individuals Living Near Inactive Uranium Mill Tailings Piles*. PB258–166, U. S. National Technical Information Service, 1976.

Tsivoglou, Ernest C., A. F. Bartsch, D. L. Rushing, and D. A. Holiday. *Report of Survey of Contamination of Surface Waters by Uranium Recovery Plants*. USPHS, Robert A. Taft Sanitary Engineering Center, Cincinnati, Ohio, 1956.

Tsivoglou, Ernest C., and R. L. O'Connell. *Waste Guide for the Uranium Milling Industry*. USPHS, Robert A. Taft Sanitary Engineering Center, Cincinnati, Ohio, 1962.

U.S. Atomic Energy Commission. *Atomic Energy and the Life Sciences (Sixth Semiannual Report)*. Washington, D.C.: Government Printing Office, 1949.

———. *Control of Radiation Hazards in the Atomic Energy Program (Eighth Semiannual Report)*. Washington, D.C.: Government Printing Office, 1950.

———. *Fourth Semiannual Report*. Washington, D.C.: Government Printing Office, 1948.

———. *Indoor Radon Daughters and Radiation Measurements in East Tennessee and Central Florida*. HASL-TM-71-8. New York: Health and Safety Laboratory, AEC, 1971.

———. *Letter from the Chairman and Members of the United States Atomic Energy Commission (Second Semiannual Report of the United States Atomic Energy Commission)*. Washington, D.C.: Government Printing Office, 1947.

———. *Letter from the Chairman and Members of the United States Atomic Energy Commission (Third Semiannual Report of the United States Atomic Energy Commission)*. Washington, D.C.: Government Printing Office, 1948.

———. *Seventh Semiannual Report of the Atomic Energy Commission*. Washington, D.C.: Government Printing Office, 1950.

———. *Theoretical Possibilities and Consequences of Major Accidents in Large Nuclear Power Plants*. Washington, D.C.: Government Printing Office, 1958.

U.S. Bureau of Mines. *Annual Report to the Secretary of the Interior for the Fiscal Year 1918*. Washington, D.C.: Government Printing Office, 1919.

———. *Bureau of Mines Bulletin No. 650*. Washington, D.C.: Government Printing Office, 1970.

U.S. Code Congressional and Administrative News 2 (1959): 2872.

U.S. Congress. House. *A Bill to Provide Federal Financial Assistance to Limit Radiation Exposure Resulting from the Use of Uranium Mill Tailings in the Area of Grand Junction, Colorado*. 92d Cong., 2d sess., 1972.

U.S. Congress. House. Committee on Armed Services. *Management of Commingled Uranium Mill Tailings: Hearings Before the Procurement and Military Nuclear Systems Subcommittee of the House Committee on Armed Services*. 97th Cong., 2d sess., 1982.

U.S. Congress. House. Committee on Interior and Insular Affairs. *H.R. Report No. 95–1480* (pts. 1, 2). 95th Cong., 2d sess., 1978.

———. Subcommittee on Energy and the Environment. *Uranium Mill Tailings Control: H.R. 13382, H.R. 12938, H.R. 12535 and H.R. 13049.* 95th Cong., 2d sess., 1978.

U.S. Congress. House. Committee on Interstate and Foreign Commerce. *H.R. Report No. 95-1480 (ii).* 95th Cong., 2d sess., 1978.

———. *Hearings on Amendments of Securities Act of 1933 (Hearings on H.R. 5701, H.R. 9319).* 84th Cong., 1st and 2d sess., July 20, 1955–May 9, 1956.

U.S. Congress. House. Committee on Mines and Mining. *Radium Hearings on H.J. Res. 185 and 186.* 63d Cong., 2d sess., January 19–28, 1914.

U.S. Congress. Joint Committee on Atomic Energy. *Atomic Development and Private Enterprise: Hearings Before the Joint Committee on Atomic Energy.* 81st Cong., 1st Sess., May 26, 1949.

———. *Atomic Power and Private Enterprise.* Joint Committee Print, JCAE, December 1952.

———. *Hearings Before the JCAE on Atomic Power Development and Private Enterprise.* 83d Cong., 1st sess., 1953.

———. *Hearings Before the Joint Committee on Atomic Energy on Federal-State Relationships in the Atomic Energy Field.* 86th Cong., 1st. sess., 1959.

———. *Hearings Before the Joint Committee on Atomic Energy on S. 3323 and H.R. 8862.* 83d Cong., 2d sess., 1954.

———. *Hearings Before the Special Subcommittee on Radiation of the Joint Committee on Atomic Energy.* 86th Cong., 1st sess., 1959.

———. *Selected Materials on Federal-State Cooperation in the Atomic Energy Field.* Washington, D.C.: Government Printing Office, 1959.

U.S. Congress. Joint Committee on Atomic Energy. Special Subcommittee on Radiation. *Hearings on Radiation Protection Criteria and Standards: Their Basis and Use.* 86th Cong., 2d sess., 1960.

U.S. Congress. Joint Committee on Atomic Energy. Subcommittee on Legislation. *ERDA Authorizing Legislation Fiscal Year 1976.* 94th Cong., 1st sess., 1975.

U.S. Congress. Joint Committee on Atomic Energy. Subcommittee on Raw Materials. *S. 2566 and H.R. 11378: Uranium Mill Tailings in the State of Utah.* 93d Cong., 2d sess., 1974.

———. *Use of Uranium Mill Tailings for Construction Purposes.* 92d Cong., 1st sess., 1971.

U.S. Congress. Joint Committee on Atomic Energy. Subcommittee on Research and Development. *Hearings on Employee Radiation Hazards and Workman's Compensation.* 86th Cong., 1st sess., 1959.

U.S. Congress. Senate. *A Bill to Provide Federal Financial Assistance to Limit Radiation Exposure Resulting from the Use of Uranium Mill Tailings in the Area of Grand Junction, Colorado.* 92d Cong., 2d sess., 1972.

———. *Senate Report No. 1211.* 79th Cong., 2d sess., 1946.

U.S. Congress. Senate. Committee on Energy and Natural Resources. *Status of the Domestic Uranium Mining and Milling Industry: Hearing on the Effects of Imports Before the Subcommittee on Energy Research and Development of the Senate Committee on Energy and Natural Resources.* 97th Cong., 1st sess., 1981.

U.S. Congress. Senate. Committee on Public Works. Subcommittee on Air and Water Pollution. *Radioactive Water Pollution in the Colorado River Basin.* 89th Cong., 2d sess., 1966.

U.S. Department of Energy. *Inactive Uranium Mill Tailing Remedial Action Program: Salt Lake City (Vitro) Site Offsite Decontamination Program Survey, Compilation of*

Candidate Properties for Remedial Action. November 18, 1983. UMTRA Project Office, Albuquerque, N.Mex.

———. Integrated Data Base for 1988: Spent Fuel and Radioactive Waste Inventories, Projections, and Characteristics. DOE/RW-0006. Washington, D.C.: Oak Ridge National Laboratory, 1988.

U.S. Department of Energy. Energy Information Administration. Domestic Uranium Mining Industry 1991 Viability Assessment. DOE/EIA-0477(91). Washington, D.C.: USDOE, 1992.

U.S. Department of Energy. Office of Environmental Management. UMTRA: Fiscal Year 1995 Annual Report to Stakeholders. Washington, D.C.: USDOE, 1995.

U.S. Department of Health, Education and Welfare. Disposition and Control of Uranium Mill Tailings Piles in the Colorado River Basin. Denver: Federal Water Pollution Control Administration, Region VII, March 1966.

U.S. Department of Health, Education and Welfare. Public Health Service. Evaluation of Radon 222 Near Uranium Tailings Piles (DER 69–1). Rockville, Md.: U.S. Department of Health, Education and Welfare, March 1969.

U.S. Department of Health, Education, and Welfare. Public Health Service (Region VIII, Denver, Colorado). Shiprock, New Mexico Uranium Mill Accident of August 22, 1960. Colorado River Basin Water Quality Control Project, January 1963.

U.S. Department of the Interior. Minerals Yearbook, 1940. Washington, D.C.: Government Printing Office, 1941.

———. Minerals Yearbook, 1941. Washington, D.C.: Government Printing Office, 1942.

———. Minerals Yearbook, 1942. Washington, D.C.: Government Printing Office, 1943.

———. Uranium Development in the San Juan Basin Region: A Report on Environmental Issues. Final Report (Fall 1990).

U.S. Department of the Interior. National Park Service. Vanadium Corporation of America Naturita Mill. Historic American Engineering Record, Report HAER CO-81.

U.S. Environmental Protection Agency. Office of Radiation Programs. Final Rule for Radon-222 Emissions From Licensed Uranium Mill Tailings. Background Information Paper, 1986.

———. Outdoor Radon Study (1974–1975): An Evaluation of Ambient Radon-222 Concentrations in Grand Junction, Colorado. Technical Note ORP/LV-77-1 (April 1977).

U.S. Federal Power Commission. National Power Survey, Environmental Research (The Report and Recommendations of the Task Force on Environmental Research to the Technical Advisory Committee on Research and Development). Washington, D.C.: Government Printing Office, 1974.

U.S. General Accounting Office. Comptroller General. Report to the Congress: Controlling the Radiation Hazard from Uranium Mill Tailings. Washington, D.C.: General Accounting Office, 1975.

———. The Uranium Mill Tailings Cleanup: Federal Leadership at Last? Washington, D.C.: General Accounting Office, 1978.

U.S. Geological Survey. Mineral Resources of the United States, Calendar Year 1901. Washington, D.C.: Government Printing Office, 1902.

———. Mineral Resources of the United States, Calendar Year 1902. Washington, D.C.: Government Printing Office, 1903.

———. Mineral Resources of the United States, Calendar Year 1903. Washington, D.C.: Government Printing Office, 1904.

———. *Mineral Resources of the United States, Calendar Year 1904*. Washington, D.C.: Government Printing Office, 1905.
———. *Mineral Resources of the United States, Calendar Year 1905*. Washington, D.C.: Government Printing Office, 1906.
———. *Mineral Resources of the United States, Calendar Year 1906*. Washington, D.C.: Government Printing Office, 1907.
———. *Mineral Resources of the United States, Calendar Year 1907*. Washington, D.C.: Government Printing Office, 1908.
———. *Mineral Resources of the United States, Calendar Year 1908*. Washington, D.C.: Government Printing Office, 1909.
———. *Mineral Resources of the United States, Calendar Year 1909*. Washington, D.C.: Government Printing Office, 1910.
———. *Mineral Resources of the United States, Calendar Year 1916*. Washington, D.C.: Government Printing Office, 1919.
———. *Twenty-first Annual Report of the United States Geological Survey, 1899–1900*. Washington, D.C.: United States Printing Office, 1901.
U.S. Nuclear Regulatory Commission. Office of Nuclear Material Safety and Safeguards. *Draft Generic Environmental Impact Statement on Uranium Milling*. April 1979.

Articles

Abbe, Robert. "Subtle Power of Radium." *Transactions of the American Surgical Association* 22 (1904): 253–62.
"Action of Radium upon the Embryo." *Scientific American Supplement* 72 (October 7, 1911): 235.
"AEC Breaks Uranium Log Jam." *Business Week* (October 11, 1958): 27–28.
"AEC Increases Price for Uranium Bearing Ores." *Engineering and Mining Journal* 149 (June 1948): 103.
Amundson, Michael A. "Home on the Range No More: The Boom and Bust of a Wyoming Uranium Mining Town, 1957–1988." *Western Historical Quarterly* 26 (1995): 483–505.
"Analyses of Green and Blue Glass from the Posilipan Mosaic." *Archaeologia* 43 (1912): 99–108.
de Alzugaray, J. Baxeres. "Manufacture and Metallurgy of Ferro-vanadium." *Mining World* 22 (June 24, 1905): 659–60.
"The Atom and Industry." *Newsweek* (November 27, 1950): 72–73.
"Atomic Age: Pure Science." *Time* (August 11, 1947): 25.
"Atomic Energy Cannot Compete as Power Source." *Science News Letter* 35 (April 8, 1939): 217.
"Atomic Energy Locked." *Literary Digest* 82 (September 20, 1924): 80–82.
"Atomic Power in Ten Years?" *Time* 25 (May 27, 1940): 44–46.
Baruch, Renee, and Madonna Ghandi. "Radioactive Waste: A Failure in Governmental Regulation." *Albany Law Review* 37 (1972): 97–134.
Berger, Rhonda S. "The Carcinogenicity of Radon." *Environmental Science and Technology* 24, no. 1 (January 1990): 30–31.
"Biggest Uranium Mill." *Time* (June 27, 1955): 80.
"The Biological Effects of Radium." *Science* 33 (June 30, 1911): 1001–5.
Bishop, H. E. "The Present Situation in the Radium Industry." *Science* 57 (March 23, 1923): 341–45.
Boaz, Mary. "Retroactive Liability for Clean-up of Hazardous Waste in *Atlas v. United*

States: The Nuclear Industry's Failed Attempt to Make the Government Pay." *Journal of Mineral and Law Policy* 6 (1990): 275–90.

"Broke and Hungry Prospector Hits Uranium Jackpot." *Business Week* (August 1, 1953): 28–30.

Byrne, John, and Steven M. Hoffman. "The Ideology of Progress and the Globalization of Nuclear Power." In *Governing the Atom: The Politics of Risk*, ed. John Byrne and Steven M. Hoffman, 11–46. New Brunswick, N.J.: Transaction, 1996.

Caley, Earle R. "The Earliest Known Use of a Material Containing Uranium." *Isis* 38, pts. 3 and 4 (1947–48): 190–93.

Cavers, David F. "The Atomic Energy Act of 1954." *Scientific American* 191 (November 1954): 31–35.

Chase, Carroll. "American Literature on Radium Therapy Prior to 1906." *American Journal of Roentgenology* 8 (1921): 766–78.

Collins, John D. "Reclamation and Groundwater Restoration in the Uranium Milling Industry: An Assessment of UMTRCA, Title II." *Journal of Natural Resources and Environmental Law* 11, no. 23 (1996): 23–98.

———. "Uranium Mine and Mill Tailings Reclamation in Wyoming—Ten Years after the Industry Collapsed." *Land and Water Law Review* 26 (1991): 489–533.

"Coming of the Giants." *Time* (May 28, 1956): 90–91.

Cook, E. "Ionizing Radiation." In *Environment*, ed. W. W. Murdock, 304. Sunderland, Mass.: Sinauer Associates, 1975.

Costello, J. M., et al. "A Review of the Environmental Impact of Mining and Milling of Radioactive Ores, Upgrading Processes, and Fabrication of Nuclear Fuels." In *Nuclear Energy and the Environment*, ed. Essam E. El-Hinnawi, 15–51. Oxford: Pergamon Press, 1980.

Crawford, Mark. "Mill Tailings: A $4 Billion Problem." *Science* 229 (August 9, 1985): 537–38.

Davidson, James M. "Vital Effects of Radium and Other Rays." *Nature* 88 (February 29, 1912): 600–602.

"Deadly Radium Gas." *Literary Digest* 101 (June 15, 1929): 19.

Dennis, L. M. "Uranium." In *The Mineral Industry: Its Statistics, Technology and Trade, 1897*, ed. Richard P. Rothwell, 653–55. New York: Scientific Publishing Company, 1898.

"Elemental Strife." *Business Week* (September 1, 1945): 101–2.

"EPA finds 'Intolerable' Radioactivity in Drinking Water Near Uranium Mines." *Environment Reporter* 6 (August 22, 1975): 651–52.

Evens, R. D. "Radium Poisoning: A Review of Present Knowledge." *Health Physics* (1980): 899–905. Reprinted from *American Journal of Public Health* 23 (October 1933): 1017–23.

"$50 Milligram Price Spurs Search for Radium." *Business Week* (September 28, 1932): 20.

Fischer, Richard P., J. C. Haff, and J. F. Rominger. "Vanadium Deposits Near Placerville, San Miguel County, Colorado." *Colorado Scientific Society Proceedings* 15 (1947): 113–34.

Fleck, Herman. "A Series of Treatises on the Rare Metals." *Proceedings of the Colorado Scientific Society*, vol. 11, 153–75. Denver: Colorado Scientific Society, 1916.

"For the Future." *Newsweek* (August 20, 1945): 59–60.

Fry R. M. "Criteria for the Long-Term Management of Uranium Mill Tailings." In *Management of Wastes from Uranium Mining and Milling: Proceedings of an*

International Symposium on the Management of Wastes From Uranium Mining and Milling, 71–83. Vienna: International Atomic Energy Agency, 1982.

"The Future of Uranium." *Time* (February 14, 1955): 94–95.

Gamson, William, and Andre Modigliani. "Media Discourse and Public Opinion on Nuclear Power." *American Journal of Sociology* 95 (1989): 1–37.

"General Mining News: Dolores County." *Engineering and Mining Journal* 68 (September 30, 1899): 406.

Gillingham, Thomas E. "Uranium." *Mining Congress Journal* 40 (February 1954): 116–18.

Goodmen, Herman. "The Romance of Radium." *Medical Journal and Record* 81 (February 19, 1930): 190–92. New York: A. R. Elliot, 1930.

Grammer, Elisa J. "The Uranium Mill Tailings Radiation Control Act of 1978 and NRC's Agreement State Program." *Natural Resources Lawyer* 13 (1981): 469–522.

Gustafson, John K. "Uranium Resources." *Scientific Monthly* 69 (August 1949): 115–20.

"Health Aspects of Radium Dial Painting." *American Journal of Public Health and the Nation's Health* 24 (April 1934): 401–2.

Hess, Frank L. "Radium, Uranium and Vanadium." In *The Mineral Industry, Its Statistics, Technology, and Trade, 1926*, ed. G. A. Roush, 601–10. New York: McGraw-Hill, 1927.

"Highlights in the History of Vanadium." *Mining and Contracting Review* 42 (July 31, 1942): 12–15.

"History's Greatest Metal Hunt." *Life* 38 (May 23, 1955): 25–35.

Hodge, Harold C. "A History of Uranium Poisoning (1824–1942)." In *Uranium, Plutonium, and Transplutonic Elements*, vol. 36 of *Handbook of Experimental Pharmacology*, ed. H. C. Hodge, J. N. Stannard, and J. B Hursh, 5–68. Berlin: Springer-Verlag, 1973.

Hohenemser, Christoph, Roger Kasperson, and Robert Kates. "The Distrust of Nuclear Power." *Science* 196 (April 1, 1977): 25–34.

Holden, Constance. "Low-Level Radiation: A High-Level Concern." *Science* 204 (April 13, 1979): 155–59.

Hollocher, Thomas C., and James J. MacKenzie. "Radiation Hazards Associated with Uranium Mill Operations." In *The Nuclear Fuel Cycle*, by the Union of Concerned Scientists, 41–69. Cambridge, Mass.: MIT Press, 1975.

Hopkins, Cyril G., and Ward H. Sachs. "Radium Fertilizer in Field Tests." *Science* 41 (May 14, 1915): 732–35.

"Hot Sands." *Newsweek* (October 18, 1971): 46.

"Hot Town." *Time* (December 20, 1971): 56.

"How to Find Uranium." *Time* (April 21, 1947): 86.

Howard, L. O. "Development of Our Radium Bearing Ores." *Salt Lake Mining Review* 15 (February 28, 1914): 13–23.

"Infinite Energy Just Out of Reach." *Literary Digest* 83 (November 15, 1924): 26–27.

Jasper, James M. "Nuclear Policy as Projection: How Policy Choices Can Create Their Own Justification." In *Governing the Atom: The Politics of Risk*, ed. John Byrne and Steven M. Hoffman, 47–66. New Brunswick, N.J.: Transaction, 1996.

"JCAE Inquiry into Mill Tailings Controversy Is Inconclusive." *Nuclear Industry* 18 (October–November 1971): 31–34.

Johnson, Jesse C. "The Romance of Uranium Mining." *Science Digest* 40 (September 1956): 58–62.

Keeney, Robert M. "The Manufacture of Ferro-alloys in the Electric Furnace." *Transactions of the American Institute of Mining Engineers* 140 (1918): 1321–73.

———. "Uranium and Vanadium." In *The Mineral Industry: Its Statistics, Technology and Trade during 1917*, ed. G. A. Roush, 720–33. New York: McGraw-Hill, 1918.

Kimball, Gordon. "Discovery of Carnotite." *Engineering and Mining Journal* 77 (June 16, 1904): 956.

Krapf, Emile F. "Recent Investigations on the Use of Radium for Malignant Diseases." *Radium* 1 (May 1913): 10–14.

Landa, Edward R. "Buried Treasure to Buried Waste: The Rise and Fall of the Radium Industry." *Colorado School of Mines Quarterly* 82 (Summer 1987): 1–77.

———. "The First Nuclear Industry." *Scientific American* 247 (November 1982): 180–93.

"Landscaping (Industrial Strength)." *Nuclear Energy* (Second Quarter 1993): 9–11.

Lang, Daniel. "A Most Valuable Accident." *New Yorker* 35 (May 2, 1959): 49–92.

Langer, R. M. "Fast New World." *Collier's* 106 (July 6, 1940): 18–55.

Laurence, William L. "The Atom Gives Up." *Saturday Evening Post* (September 7, 1940): 12–63.

Lind, S. C. "Radium Production in America." *Chemical and Metallurgical Engineering* 26 (May 31, 1922): 1012–13.

Lorenz, Egon. "Radioactivity and Lung Cancer: A Critical Review of Lung Cancer in the Miners of Schneeberg and Joachimsthal." *Journal of the National Cancer Institute* 5 (August 1944): 1–15.

Lounsbury, James E. "Famous Pittsburgh Industries: The Standard Chemical Company of Pittsburgh, Pa." *Crucible* 22 (June 1938): 134–37.

MacArthur, John S. "The Radium Industry and Reconstruction." *Engineering and Mining Journal* 107 (April 5, 1919): 605–6.

Magee, John. "The Uranium Mill Tailings Radiation Control Act of 1978." *Ecology Law Quarterly* 8 (1980): 801–9.

Makhijani, Arjun, Stephen I. Schwartz, and William J. Weida, "Nuclear Waste Management and Environmental Remediation," In *Atomic Audit: The Costs and Consequences of U. S. Nuclear Weapons since 1940*, ed. Steven I. Schwartz, 353–94. Washington, D.C.: Brookings Institution Press, 1998.

"Malignant Growths Resulting from Exposure to Radioactive Substances." *American Journal of Public Health and the Nation's Health* 22 (July 1932): 760–61.

"Management of Inactive Uranium Mill Tailings." *Journal of Environmental Engineering* 112 (June 1986): 490–537.

McKee, Thomas M. "Early Discovery of Uranium Ore in Colorado." *Colorado Magazine* 32 (July 1955): 191–203.

Metzger, H. Peter. "AEC vs. The Public: The Case of the Uranium Tailings." *Science News* 106 (July 13, 1974): 31.

Meyer, Larry L. "The Time of the Great Fever: U-Boom on the Colorado Plateau." *American Heritage* 32 (June–July 1981): 74–80.

Meyers, Burt. "Uranium Jackpot." *Engineering and Mining Journal* 154 (September 1953): 72–75.

Meyers, Harold B. "The Great Uranium Glut." *Fortune* 69 (February 1964): 108–64.

Miller, Byron S. "A Law Is Passed—The Atomic Energy Act of 1946." *University of Chicago Law Review* 15 (Summer 1948): 799–829.

Miller, S. E., et al. "Health Protection of Uranium Miners and Millers." *American Medical Association Archives of Industrial Health* 14 (1956): 48–55.

Millikan, Robert A. "The Significance of Radium." *Bulletin of the California Institute of Technology* 29 (June 1921): 3–21.
Moore, Richard B. "Radium." In *The Mineral Industry: Its Statistics, Technology and Trade during 1920*, ed. G. A. Roush, 615–19. New York: McGraw-Hill, 1921.
———. "Radium." In *The Mineral Industry, Its Statistics, Technology, and Trade, 1922*, ed. G. A. Roush, 617–20. New York: McGraw-Hill, 1923.
———. "Radium Production." *Science* 49 (June 13, 1919): 564–66.
———. "Uranium and Vanadium." In *The Mineral Industry: Its Statistics, Technology and Trade during 1920*, ed. G. A. Roush, 705–12. New York: McGraw-Hill, 1921.
———. "Uranium and Vanadium." In *The Mineral Industry: Its Statistics, Technology and Trade, 1922*, ed. G. A. Roush, 711–17. New York: McGraw-Hill, 1923.
National Bureau of Standards. Committee on Radiation Protection and Measurements. "Somatic Radiation Dose for the General Population." *Science* 131 (February 1960): 482–86.
"New Mills Break Up Uranium Bottleneck." *Business Week* (August 6, 1955): 102–5.
Newman, James R. "The Atomic Energy Industry: An Experiment in Hybridization." *Yale Law Journal* 60 (December 1951): 1263–1394.
Niehoff, Richard O. "Organization and Administration of the United States Atomic Energy Commission." *Public Administration Review* 8 (May 1948): 91–102.
Novick, Sheldon. "Radioactive Mining Wastes." *Scientist and Citizen* 8 (August 1966): 10–12.
"NRC Radon Impact Estimates 'Grossly' in Error, Says Ex-Oak Ridge Official." *Environment Reporter* 8 (November 25, 1977): 1139–40.
"Nuclear Waste Kills Investment." *Business Week* (March 17, 1980): 41–43.
O'Neill, John J. "Enter Atomic Power." *Harper's Magazine* 181 (June 1940): 1–10.
Oppenheimer, J. Robert. "The Atom Bomb as a Great Force for Peace." *New York Times Magazine* (June 9, 1946): 7–60.
Owen, Anthony D. "Short-Term Price Formation in the U.S. Uranium Market." *Energy Journal* 6 (1985): 37–49.
Parsons, A. T. "Radium, with Special Reference to Luminous Paint." *Journal of the Oil and Colour Chemists' Association* 12 (January 1929): 3–28.
Parsons, Charles L. "Our Radium Resources." *Science* 38 (October 31, 1913): 612–20.
Patterson, Walter C. "Chernobyl—The Official Story." *Bulletin of the Atomic Scientists* 42 (November 1986): 34–36.
Peller, Sigismund. "Lung Cancer among Mine Workers in Joachimsthal." *Human Biology* 11 (1939): 130–43.
"Pennies for Uranium." *Time* (April 5, 1954): 89–90.
"Plenty of Uranium." *Science News Letter* 54 (December 25, 1948): 403.
Plough, Alonzo P., and Sheldon Krimsky. "The Emergence of Risk Communication Studies: Social and Political Context." *Science, Technology & Human Values* 12 (Summer 1987): 4–10.
Pope, Elizabeth. "The Richest Town in the U.S.A." *McCall's* (December 1956): 38–39, 99–104.
"Private Development of Uranium Is Encouraged." *Engineering and Mining Journal* 149 (June 1948): 67.
"Punctured Boom." *Business Week* (December 17, 1955): 124–26.
"Pure Science." *Time* (August 11, 1947): 25.
Pursey, William A. "Biological Effects of Radium." *Science* 33 (June 30, 1911): 1001–5.

———. "Effects of Radium on Living Cells." *Scientific American*, Suppl. 22 (July 22, 1911): 56–61.

"Putting the Atom to Work." *Popular Mechanics* (May 1938): 690–93, 116A–119A.

"Radium in the Home." *Literary Digest* 82 (September 13, 1924): 78–81.

"Radium Victim No. 41." *Life* 31 (December 17, 1951): 81.

"Radon? Sure. So What Else Is New? Ask the Folks at Grand Junction." *Nuclear Industry* 18 (October–November 1971): 34–39.

Ramsey, R. R. "Radium Fertilizer." *Science* 42 (August 13, 1915): 219.

Rapoport, Rodger. "The Trouble with 90.5 Million Tons of Radioactive Tailings." *Los Angeles Times West Magazine* (April 12, 1970): 10–15.

"Recent Contributions to Radium Therapy, III." *Radium* 4 (October 1914): 74–80.

"Regulation of Uranium Mill Tailings; Uranium Mill Tailings Radiation Control Act of 1978." In *Treatise on Environmental Law*, ed. Frank P. Grad, 4:6–124–6–134. New York: Matthew Bender & Co., 1992.

"Report Says Radon Exposure Major Hazard in Living Near Uranium Tailings Deposits." *Environment Reporter* 7 (May 14, 1976): 49–50.

"Residential Radioactivity: Sharing Responsibility." *Science News* 100 (December 11, 1971): 390.

Riccitiello, David, et al. "Uranium Mining and Milling." *Workbook* 4 (1979): 222–35.

Rutherford, Ernest, and Frederick Soddy. "Radioactive Change." *London, Edinburgh, and Dublin Philosophical Magazine and Journal of Science* 6 (1903): 576–91.

Savage, Wallace. "Radioactive Luminous Materials." *Chemical and Metallurgical Engineering* 19 (September 28, 1918): 515–17.

Scott, Elizabeth V. "Unfinished Business: The Regulation of Uranium Mining and Milling." *University of Richmond Law Review* (1984): 615–53.

"SEC Wars on Uranium, Oil Promotions." *Business Week* (July 23, 1955): 66.

Shuey, Chris. "Bringing Tailings under Control." *Workbook* 10 (1985): 110–13.

Silver, Ken. "The Yellowed Archives of Yellowcake," *Public Health Reports* 111 (March–April 1996): 116–27.

Sorensen, Don. "Wonder Mineral: Utah's Uranium." *Utah Historical Quarterly* 31 (Summer 1963): 280–90.

Strauss, Jerome. "Radium, Uranium, and Vanadium." In *The Mineral Industry: Its Statistics, Technology, and Trade during 1939*, ed. G. A. Roush, 519–26. New York: McGraw-Hill, 1940.

———. "Radium, Uranium, and Vanadium." In *The Mineral Industry: Its Statistics, Technology, and Trade during 1940*, ed. G. A. Roush, 541–49. New York: McGraw-Hill, 1941.

Stryker, Perrin. "The Great Uranium Rush." *Fortune* 50 (August 1954): 89–158.

"Surgeon General Sets Radiation Levels for Buildings Constructed with Uranium." *Environment Reporter* 1 (August 21, 1970): 446–47.

Taylor, Peter J. "Technocratic Optimism, H. T. Odum, and the Partial Transformation of Ecological Metaphor after World War II." *Journal of the History of Biology* 21 (1988): 213–44.

Teeple, David S. "The Coming Uranium Bust." *American Mercury* 81 (September 1955): 34–40.

"U_3O_8: Energy from Wyoming's Powerful Sand." *In Wyoming* 13 (February–March 1980): 48–57.

"United States Mineral Production in 1897." *Engineering and Mining Journal* 65 (May 28, 1898): 638.

"Uranium, 1956." *True West* 3 (May–June 1956): 4–35.
"Uranium Grows Up—Big Business Now." *U.S. News and World Report* (April 6, 1956): 90–92.
"Uranium Industry Leaps to Maturity." *Business Week* (April 21, 1956): 30–32.
"Uranium Is Too Scarce for Use as Source of Fuel." *Science News Letter* 53 (May 1, 1948): 283–84.
"Uranium Jackpot." *Time* (September 30, 1957): 89.
"Uranium Legacy." *Workbook* 8 (November–December 1983): 192–207.
"Uranium Mill Tailings: Congress Addresses a Long-Neglected Problem." *Science* 202 (October 13, 1978): 191–202.
"Uranium Mystery." *New Republic* 154 (April 16, 1966): 36–37.
"Uranium Mystery in the Colorado Basin." *New Republic* 154 (March 5, 1966): 9.
"U.S. Atomic Energy Commission Announces Program to Stimulate Production of Domestic Uranium." *Engineering and Mining Journal* 149 (May 1948): 108.
"Vanadium Quiz." *Business Week* (July 14, 1945): 36–38.
Viol, Charles H. "Radium Production." *Science* 49 (March 7, 1919): 227–28.
Viol, Charles H., and Glenn D. Krammer. "The Application of Radium in Warfare." *Transactions of the American Electrochemical Society* 32 (1918): 381–90.
Wagar, W. Warren. "Toward a World Set Free: The Vision of H. G. Wells." *Futurist* 17 (1983): 24–31.
Watford, Glen A., and John A. Wethington, Jr. "Radiological Hazards of Uranium Mill Tailings Piles." *Nuclear Technology* 53, no. 3 (June 1981): 295–301.
"What Is the Atom's Industrial Future?" *Business Week* (March 8, 1947): 21–30.
"When H. G. Wells Split the Atom: A 1914 Preview of 1945." *The Nation* 161 (August 18, 1945): 154–56.
"Why Buy Uranium Stocks Now?" *Newsweek* (August 29, 1955): 57–58.
Wickham, Louis, and Paul Degrais. "Radium: Its Uses in Cancer and Other Diseases." *Contemporary* 98 (August 1910): 174–88.
Wilson, Ellen. "Some Like it Hot." *Environmental Action* 17 (November–December 1985): 27–32.
Wood, Nancy. "America's Most Radioactive City." *McCall's* 97 (September 1970): 46–50, 122.
Wylie, Philip. "Deliverance or Doom." *Collier's* 116 (September 29, 1945): 18–80.

Archives

Wayne Aspinall Papers. Penrose Library, University of Denver, Denver, Colorado.
Colorado Historical Society. State Department of Health File, 1967–68. Denver.
Colorado State Archives, Denver.
Frank E. Moss Papers. Marriott Library Special Collections, University of Utah, Salt Lake City.
MX Information Center Papers. Marriott Library Special Collections, University of Utah, Salt Lake City.
New Mexico Environmental Department, Hazardous Waste Bureau, Mixed Waste Section. Santa Fe.
Southwest Research and Information Center. Albuquerque, New Mexico.
UMTRA Project Office Archives. Albuquerque, New Mexico.
Union Carbide Company Archives. Grand Junction, Colorado.
U.S. Bureau of Mines. Record Group 70, General Classified File, General Correspondence File. National Archives and Records Administration, Washington, D.C.

U.S. Congress. Joint Committee on Atomic Energy. Record Group 128, General Correspondence File. National Archives and Records Administration, Washington, D.C.

U.S. Department of Energy. Atomic Energy Commission, Glenn Seaborg Collection. USDOE Archives, Germantown, Maryland.

U.S. Department of Energy. Atomic Energy Commission, Secretariat File. USDOE Archives, Germantown, Maryland.

U.S. Nuclear Regulatory Commission. Public Documents Reading Room, Washington, D.C.

U.S. Public Health Service. National Archives and Records Administration, Washington, D.C.

Utah Department of Environmental Quality. Vitro Case Records. Salt Lake City.

Dissertations and Theses

Pittman, Francis L. "The Direct Production of Uranium Steel." M.S. thesis, Colorado School of Mines, 1914.

Shumway, Gary Lee. "A History of the Uranium Industry on the Colorado Plateau." Ph.D. diss., University of Southern California, 1970.

Newspapers

Albuquerque Journal, 1989
Ann Arbor News, 1994
(Grants, N.Mex.) *Cibola County Beacon*, 1958
(Grand Junction, Colo.) *Daily Sentinel*, 1966–1998
Denver Post, 1954–71
Deseret News, 1998
Durango Herald, 1966
Durango Herald-Democrat, 1948
Fruita (Colo.) *Times*, 1982
High Country News, 1994–97
Las Vegas Sun, 1998
Los Angeles Times West Magazine, 1970
New York Times, 1933–71
New York World, 1928
Pittsburgh Press, 1980
Rocky Mountain News, 1965–71
Salt Lake Tribune, 1998
(Moab, Utah) *Times-Independent*, 1954
Washington Daily News, 1966

Index

air pollution: in mines and mills, 103, 104–7; from tailings, 107–15
Ambrosia Lake, New Mexico, 153, 162, 163–64
American Rare Metal Mining and Manufacturing Company, 21, 67
Anderson, Clinton P., 47
Animas River, New Mexico: pollution of, 77–78, 85
Animas River Conference, 85, 87
Aspinall, Wayne, 100, 110–11, 129–30, 136
Atlas Minerals Company, 98, 217–18n. 46, 220n. 19
Atomic Energy Act: 13, 32–33, 43, 48, 57, 79; Agreement Program, 137–38, 156; 1954 amendment to, 13, 44–46, 48, 80, 176; 1959 amendment to, 137–38
Atomic Energy Commission (AEC): accepts responsibility for tailings pollution, 142–43; agrees to assist in indoor radiation evaluation, 130, 132; collaborates with USPHS on radon testing, 110; commitment to private atomic energy programs by, 41–46, 55–58; creation of, 32–33; denies liability for tailings pollution, 136–42, 143–44; differences with FWPCA of, 93, 94–95, 97–98; downplays tailings hazards, 5–6, 12–13, 14–15, 107; fails to support remedial action on tailings used in construction, 134–36; governing of milling industry by, 11–12, 83–84; initial challenges of, 39–40; joint statement on tailings management, 96–97, 99, 102–3; minimizes nuclear risks, 3–4, 75, 78–79; ore-buying program of, 58–60, 193n. 23; perceived lack of concern for health issues of, 12, 115–16; on radium, 78–81, 91; regulatory threshold of, 79, 89, 107; relations with JCAE of, 13–14, 47; reluctant to regulate abandoned tailings, 88–89; reluctant to regulate mills, 85–87; resists study of radon in buildings, 123–24, 127, 129; secrecy of, 40, 49; water pollution and, 73–88

Ball, Howard, 84
Balogh, Brian, 5
Bardin, B. J., 108
Barker, Robert, 87
Batie, Ralph, 105, 106
Boggs, Larry, 166
Boyer, Paul, 2
Burwell, Blair, 55

237

Canonsburg, Pennsylvania, 206n. 4
carnotite, 20, 21
Chernobyl, 5
Cleere, Roy, 129, 141
Climax Uranium Company mill, Grand Junction, Colorado, 106, 120–21, 122, 123, 139
Cohn, Steven Mark, 4
Cold War, 2, 13, 42, 49, 60
Colorado Department of Public Health: air quality tests by, 109; orders stop to release of tailings, 122–23; radiation survey of buildings by, 124–25, 129; regulation of tailings by, 99, 102; on responsibility for tailings, 111–12
Colorado State Board of Health, 98, 99, 107
Conference in the Matter of Pollution of Interstate Waters of the Colorado River and Its Waters, 111
Curie, Pierre and Marie, 20

Defense Minerals Exploration Administration, 62–63
Department of Energy, 155, 172
Department of Health, Education, and Welfare, 91, 97
Deutsche Roentgen Gesellschaft, 28
Dominick, Peter, 133, 136
Drinkwater, Terry, 94
Durango, Colorado mill, 69, 77–78

Edwards, Page, 98
Einstein, Albert, 35
Eisenhower, Dwight D., 41
Environmental Protection Agency (EPA), 128, 131, 135; on abandoned tailings, 151; during Reagan administration, 167; standards for tailings remediation, 159, 171, 172–73; surveys gamma radiation in Salt Lake County, 148; tests water in New Mexico, 153

Fain, Gerald, 132
Faulkner, Rafford, 108
Federal Radiation Council, 130, 131
Federal Water Pollution Control Administration (FWPCA): differences with AEC, 93, 94–95, 97–98; joint statement on tailings management, 96–97, 99, 102–3; report on tailings pollution of, 90–94, 96, 119
Fermi, Enrico, 40
Furry, Dean, 108

gamma radiation, 121, 210n. 34
General Accounting Office, 155
Gmelin, Christian Gottlob, 18–19
Gordon, Charles, 108
Grand Junction, Colorado: cost of remedial action in, 136; difficulties of cleanup in, 164–65; evidence of health problems in, 127–28; negative publicity about, 125, 126; radiation pollution of buildings in, 121, 123, 124–25, 129, 130; use of tailings in construction in, 120–21
Grants, New Mexico, 73, 162–64
Grants Mineral Belt, 153
Groves, Leslie, 39, 52
Gustafson, John K., 55, 58

Handler, Philip, 182n. 9
Hennessey, Joseph, 81–82
Holden, Constance, 182n. 9
Hollis, Roy, 98
Homestake-New Mexico Partners mill, Grants, New Mexico, 82

Interagency Steering Committee (on radiation pollution of buildings), 131–32, 133–34
Interagency Technical Committee on Control of Uranium Mill Tailings Piles, 113
International Commission on Radiation Protection, 78, 130

Jasper, James M., 48
Joe Junior mill, Colorado, 67, 68
Johnson, Edwin, 59
Johnson, Jesse C., 58, 60

Joint Committee on Atomic Energy (JCAE), 3; creation of, 33–34; 1953 hearings of, 42–43, 44; relations with AEC of, 13–14, 47
Jordan, Walter H., 153–54

Keller, Glen, 141
Kempe, C. Henry, 127–28
Kerr-McGee Corporation, 98, 109, 163, 197n. 5
Klaproth, Martin Heinrich, 18
Kusnetz, Howard L., 125

Lee, Philip, 97, 109
leukemia, 7, 209–10n. 28
Lilienthal, David E., 37, 38, 40, 56
Lind, Samuel, 29
Little, A. O., 92
Liverman, James, 150
Love, John, 127
lung cancer: from radon, 9–10, 103, 104, 122

Makhijani, Arjun, 14
Manhattan Engineer District (Manhattan Project), 31, 39, 52–53, 69
Marks, Herbert, 56–57
McMahon, Brian, 42
Metals Reserve Corporation (MRC), 51–52, 69
Metzger, H. Peter, 10
Miller, Byron, 40, 54
Millikan, Robert A., 53–54
Milliken, Eugene, 59
Monticello, Utah mill, 88
Morris, Peter, 95, 97, 118
Moss, Frank, 149, 151, 152
Muskie, Edmund S., 94, 95, 96

Nader, Ralph, 10
National Committee for Radiation Protection, 78
Naturita, Colorado mill, 68, 69, 77
Newman, James, 41, 54, 190n. 46
New Mexico Department of Public Health, 85
Newmire, Colorado mill, 68
New Republic, 93, 201n. 47
Niehoff, Robert, 49

Nininger, Robert, 57
nuclear energy/power: civilian, 39–46; consequences of accidents from, 191n. 48; critics of, 4–5, 74; early minimization of risks of, 3–4; early speculation about, 34–36; optimistic predictions about, 2, 3, 37–38; wariness of, 2–3
Nuclear Regulatory Commission, 154, 159, 170–71, 173

Oppenheimer, J. Robert, 38, 40
Owens, Wayne, 149

Palos Park, Illinois, 206n. 4
Pendleton, Robert, 148
Peterson, Earl, 119
pitchblende, 20, 21
polonium 214 and 218, 8, 9
Poulot, Charles, 21, 67
President's Commission on National Goals, 117
Price-Anderson Act, 45, 191n. 48

radiation exposure: guidelines, 109, 130–31, 210n. 34; threshold, 78
radium: AEC argues not responsible for, 78–81; AEC claims not health hazard, 91; dangers of, 27–29; discovery of, 20; end of U. S. industry, 26–27; refineries in U. S. of, 24; therapeutic uses of, 17–18, 22–23, 24, 27–28; use in paint of, 24–25; water pollution from uranium tailings by, 76–78;
Radium Luminous Materials Corporation, 26
radon: as cause of lung disease, 9–10, 103, 104, 122; danger to mill workers from, 104; decay of, 8–9, 121–22; emitted from tailings piles, 104, 109–10; measurement of, 208n. 17; not considered health hazard, 113–15; risks from, 154; from tailings used in construction, 122
Rampton, Calvin, 149–50
Reagan, Ronald, 166–67
Rennels, Duane, 155–56
Rolph, Elizabeth, 75
Romans: use of uranium by, 17, 18

Index   239

Ross, Robert, Jr., 127–28
Rutherford, Sir Ernest, 34–35

Salt Lake City, Utah. *See* Vitro Chemical Company
Schlesinger, James, 142
Schwartz, Stephen I., 14
scientists: as molders of opinion, 37–38
Seaborg, Glenn, 89, 97, 127, 133
secrecy, 40, 49
Senate Subcommittee on Air and Water Pollution, 94–96, 118
Siek, Robert, 121
Simon, Herbert, 75
Smith, Paul, 151
Smyth, Henry D., 31
Snelling, Robert, 121
Soddy, Frederick, 34–35
Standard Chemical Company, 24, 25, 26, 27, 67
Steen, Charlie, 61, 64, 71
Stein, Murry, 95
Steinfeld, Jesse L., 130–31
Stout, William, 38
Strauss, Lewis, 2
Students for a Democratic Society, 2
Study of Inactive Uranium Mill Sites and Tailings Piles, 151, 152–53

tailings: AEC downplays hazards of, 5–6, 12–13, 14–15; air pollution from, 107–15; cost of regulating, 178; creation of, 6–7, 75–76; dam failures, 76, 197n. 5; early methods of disposal of, 76; health danger of, 7–10, 75; mill company interest in management of, 98; pile stabilization, 201n. 52; radium from, 76–77; radon from, 104, 109–10; reduction of pollution from active piles, 118; reprocessing of, 155–56; significance of debate about, 15–16; use in construction of, 119–44; water pollution from abandoned, 88–101. *See also* Uranium Mill Tailings Radiation Control Act; Uranium Mill Tailings Remedial Action Project

Tailings for Construction Hearing, 133–41
technocrats, 3, 181n. 6
Three Mile Island, 5, 163

Union Mines Development Corporation, 53
U.S. Public Health Service: evaluates tailings pollution, 77–78, 87, 113; indoor radiation pollution and, 123, 128–29, 130; joint statement on tailings management, 96–97, 99, 102–3; radon testing by, 109, 110; report on use of tailings in construction, 140
United States Vanadium, 53, 68, 196n. 44
uranium: AEC ore-buying program for, 58–60, 61; discovery of, 18; early assessment of U. S. reserves, 53–54, 55; early boom in, 21–22; early glut of, 25–26; early mining of, 20, 22–23; early reliance on foreign sources of, 52–53, 54–55, 59; loss of federal price support for, 162–63, 167–69; 1950s boom in, 61–66, 194n. 33; price of, 58, 168, 169; production from vanadium mining tailings of, 69; reassessment of U. S. reserves of, 60–61; therapeutic uses of, 19; use in glass of, 17, 18, 19; use in World War II of, 51, 52, 69
uranium fuel cycle, 6–7, 75
uranium mills: closure of, 163, 169; construction of new, 65, 71; conversion of vanadium mills to, 70; dangers to workers at, 104–7; early U. S., 21, 24, 67; process at, 6–7; reluctance to invest in, 70–71. *See also* tailings; names of individual mills
Uranium Mill Tailings Control Hearing, 154–56
Uranium Mill Tailings Radiation Control Act (UMTRCA), 10, 157–61, 179; extensions of, 173, 174; Ground Water Project, 173, 174; lassitude in carrying out of, 175–76; legal challenges to, 170–72; unfortunate timing of, 166–67, 169

Uranium Mill Tailings Remedial Action Project (UMTRA), 1, 159, 172–74
uranium 234, 8
uranium 235, 51
uranium 238, 7, 8, 51
Uravan, Colorado mill, 69, 85
Utex mill, Moab, Utah, 71, 82

vanadium, 22, 51–52, 67–70
Vanadium Corporation of America, 52, 53, 68, 108, 196n. 44
Viscusi, W. Kip, 220n. 25
Vitro Chemical Company: tailings of, 145–52
Vitro Rare Metals Company, 206n. 4
Voilleque, Charles, 21, 67

Walker, J. Samuel, 10
Walters, Jay, Jr., 63
water pollution: AEC adopts mill licensing to control, 83–84, 87–88; AEC avoids responsibility for, 78–83; AEC sees as not immediate hazard, 85–87; Colorado regulations to reduce, 99; in Colorado River Basin, 76–78, 85, 90, 93;
Weida, William J., 14
Wells, H. G., 34
Westinghouse Electric Company, 2
Wigton, Chester, 108
Wilson, Carroll, 70
Wolf, Bernie, 106–7